Schering Foundation Workshop 2
Round Table Discussion on
BIOSCIENCE ⇌ SOCIETY

Schering Foundation Workshop

Series Editors: Günter Stock
Ursula F. Habenicht

Vol. 1
Bioscience ⇌ Society
Workshop Report
Editors: D.J. Roy, B.E. Wynne, R.W. Old

Vol. 2
Round Table Discussion on Bioscience ⇌ Society
Editor: J.J. Cherfas

This series will be available on request by
Schering Research Foundation,
P.O. Box 65 03 11, W–1000 Berlin 65, FRG

Held and published on behalf of
Schering AG, Berlin

This Round Table Discussion
was sponsored by:
Senate of the City of Berlin
Schering AG, Berlin

Schering Foundation Workshop 2

Round Table Discussion on BIOSCIENCE ⇌ SOCIETY

J.J. Cherfas, Editor

Report of the Round Table Discussion on
BIOSCIENCE ⇌ SOCIETY
Berlin, 1990, December 1

Moderator: D.J. Roy
Lectures: H.T. Engelhardt Jr., P. Kafka, M.N. Maxey, M. McGregor
Panel: J.J. Cherfas, E. Fox Keller, D. Rössler, E.-L. Winnacker
Consultant Organizer: S. Bernhard

Springer-Verlag
Berlin Heidelberg GmbH

Text Preparation: P. Knowlton, D.L. Lewis

Photographs: Lohmann, Schering AG

ISBN 978-3-662-22389-5   ISBN 978-3-662-22387-1 (eBook)
DOI 10.1007/978-3-662-22387-1

This work is subject to copyright. All rights are reserved, whether the whole or part of the material is concerned, specifically the rights of translation, reprinting, reuse of illustrations, recitation, broadcasting, reproduction on microfilms or in other ways, and storage in data banks. Duplication of this publication or parts thereof is permitted only under the provisions of the German Copyright Law of September 9, 1965, in its current version, and permission for use must always be obtained from Springer-Verlag Berlin Heidelberg GmbH. Violations are liable for prosecution under the German Copyright Law.

© Springer-Verlag Berlin Heidelberg 1991
Originally published by Springer-Verlag Berlin Heidelberg New York in 1991
Softcover reprint of the hardcover 1st edition 1991

Ths use of general descriptive names, registered names, trademarks, etc. in this publication does not imply, even in the absence of a specific statement, that such names are exempt from the relevant protective laws and regulations and therefore free for general use.

Product liability: The publishers cannot guarantee the accuracy of any information about dosage and application contained in this book. In every individual case the user must check such information by consulting the relevant literature.

Typesetting with TEX: Danny Lee Lewis Buchproduction, Berlin
Reproduction of the Figures: Gustav Dreher GmbH, Stuttgart

21/3130-5 4 3 2 1 0 – Printed on acid-free paper

Table of Contents

Foreword
G. Stock   1

Welcoming Address
B. Riedmüller   3

Introductory Remarks
G. Vita   7

Introduction
D.J. Roy   11

Does Bioscience Threaten Human Integrity?
M. McGregor   15

Discussion   19

Does Bioscience Threaten Ecological Integrity?
P. Kafka   29

Discussion   35

What's Wrong with the Interaction Between Bioscience and Society?
H.T. Engelhardt, Jr.   43

Discussion   47

What Actions Are Required to Improve the Present Uneasy Relationship Between Bioscience and Society?
M.N. Maxey   57

Discussion	63
Closing Remarks *G. Stock*	73
Foto/CV's of Lecturers/Panel Members	77
List of Participants	89

Foreword

To mark the occasion of the 100th anniversary of Schering research **two** events took place:

a) a **Workshop on Bioscience ⇌ Society,** which was held November 24–30, 1990 in Berlin.

The goal of the workshop was:

> Do current and anticipated developments in bioscience require a new covenant between science and society?

The results have been published as the first volume of a newly created "Schering Foundation Workshop" series* by John Wiley & Sons Ltd, Chichester, England (D.J. Roy, E.W. Wynne, and R.W. Old, eds. (1991), *Bioscience ⇌ Society*, Chichester: John Wiley & Sons);

b) a **Round Table Discussion,** held on December 1, 1990, where the results of the workshop were discussed by a special invited panel.

The meetings were sponsored by the Senate of the City of Berlin and Schering AG, Berlin.

This publication contains summary reports of the four discussion groups of the workshop presented by their moderators, and the comments of the panel at the Round Table Discussion.

This is the second volume in the "Schering Foundation Workshop" series. We hope that this series will contribute to a better understanding of science and to a better recognition of its merits.

G. Stock

---

* This series will be available on request from "Schering Research Workshop", Müllerstrasse 178, W-1000 Berlin 65.

From left to right:
G. Vita, B. Riedmüller, G. Stock. S. Bernhard, M. McGregor, P. Kafka, H.T. Engelhardt, Jr., M.N. Maxey

Welcoming Address

Barbara Riedmüller*:
Ladies and Gentlemen, I bring the greetings and best wishes of the Government of Berlin to Schering AG on the occasion of the 100th anniversary of Schering research, in particular from Governing Mayor Diepgen, who regrets that other duties prevent his attending today.

As the Senator responsible for science and research in Berlin, I would like to link these good wishes for the continued successful development of research at Schering with the government's thanks to the company for its steadfast commitment and socially responsible cooperation with Berlin.

A visible sign of this fruitful and successful cooperation in recent times is the joint foundation in this city of the Institut für Genbiologische Forschung, which is one of five genetic engineering research centers in Germany. The availability of generous funds for establishing foundation professorships and supporting other joint scientific activities in tertiary educational institutions in Berlin is also characteristic of this cooperation.

Schering AG was one of the few large enterprises in Berlin to maintain its head office here after the Second World War, not only in name but also in its day-to-day business. In doing so, this great enterprise not only helped Berlin as a patron, but also supported the city during difficult times in its reconstruction, through investment, creating jobs and, not least, by paying taxes. This will not be forgotten.

Today Berlin is one of the main scientific centers in Germany and in Europe. The political changes and the new circumstances resulting from German unity have now placed Berlin in the rather ungratifying position of "executor of estate" for scientific research in what used to be East Germany. The consequences of the new task also involve the risk of placing new burdens on what has already been achieved.

---

* Former Senator for Science and Research in Berlin

At the same time we also recognize the many opportunities that the research potential concentrated in the eastern part of our city offers. I remind you that as of October 3, 1990 Berlin has three universities, three university teaching hospitals in addition to the two existing specialized colleges, two further technical colleges, three schools of the arts, as well as several scientific and artistic institutions.

In addition, decisions are to be made on the future of more than half of all the institutes of the former GDR Academy of Sciences, which are concentrated in Berlin.

The transition of the research institutes into an independent expert system will require a balance of state promotion of science on the one hand and – thanks to our public budget – the need to shrink on the other, always bearing in mind the social feasibility of such shrinkage. Nevertheless, there does now exist a rare opportunity to build quickly in Berlin a center of science of European dimensions.

For various reasons Berlin must not lose this unique opportunity, which will be so significant for the future. It will help to retain the human capital of Berlin and the region, and those people will decisively influence cultural and economic development. It will also offer an opportunity to build up new fields of research for an ecological restructuring of industrial society.

It will be difficult. There is not enough money, but added to this is the city's special situation; Berlin is neglected by the German Federal Government and seen as a rival by the other states. But we cannot behave like an old state in West Germany or proceed on the assumption of a guaranteed income, because Berlin's territory and population have almost doubled and new structures must be built up everywhere in the newly-added parts. One could say that the process of German unification is taking place in miniature in Berlin, without even a hint of solidarity manifested by the old West German states.

In all decisions in the field of science, but particularly in research, limited funds must not lead to a decline in quality. Politics must find new ways to make investments possible in the future. We are therefore grateful when industry also commits itself in the other part of the city, and we have noted with pleasure that Schering, too, has clearly demonstrated its interest in the future of research institutions by commissioning research projects. In doing so Schering is making an additional contribution to the improvement of equipment supplies and thus to the performance of the institutes themselves. In the present situation this provides important psychological support for the people who work there.

In view of this exemplary behavior of a large chemical and pharmaceutical enterprise, it is almost taken for granted when Schering, in celebrating an

Welcoming Address

outstanding event like the 100th anniversary of its research tradition, not only dedicates itself to the newest developments of bioscience but also consciously places this topic in conjunction with society.

I would therefore like to congratulate Schering AG, its board, and all those involved in these anniversary festivities for celebrating with a one-week workshop on "Bioscience $\rightleftharpoons$ Society" and for organizing the Round Table Discussions to explore the results of that workshop.

Introductory Remarks

Guiseppe Vita*:
Thank you, Dr. Riedmüller, for your welcoming address and for coming here today, the day before a very important Sunday in the history of a new Germany.

Ladies and Gentlemen, I wish to extend a cordial welcome to all of you on behalf of Schering. I thank you for accepting our invitation, particularly because we are not celebrating our research centenary at Schering with the usual type of ceremony; this is more of a working party.

To celebrate a centenary of research may seem to some of you like old men reminiscing about the glorious past. To us, it is looking at a hundred years of successful work in developing pioneering drugs, especially hormones, and contrast media. To us it means a hundred years of innovation, of moving forward, breaking new ground time and time again. We have coupled innovative spirit with tradition and experience, and for all our years we still feel young.

Our history of innovation began back in 1890, when Schering AG developed its first scientifically-researched drug, Piperazine, a preparation against rheumatism and gout. We made a name for ourselves in 1930 with the very first water-soluble X-ray contrast medium, Uroselectan, a diagnostic aid for visualizing the kidneys and the urinary tract. We followed this with various innovative contrast media for use in a whole range of indications. The very latest brainchild of Schering's diagnostic research is the world's first contrast agent for magnetic resonance imaging, Magnevist, which came out in 1988.

The second field of research of paramount importance to Schering has for many decades now been hormones. Schering cooperated with the Nobel Prize winner Professor Adolf Butenandt, Director of the Kaiser-Wilhelm-Institut, the precursor of today's Max-Planck-Institut. In the 1930s, this cooperation led to our first hormone preparation for replacement therapy. As you will know, Schering gained further recognition with its introduction, in 1960,

---

* Chairman of the Board of Directors of Schering AG

of Europe's first "pill". Ten years later Schering was first to develop an antihormone. I am happy to see that one of our pioneers in steroid hormone research is with us here today, Professor Russel Marker.

Our success as a research-based company is to a large extent due to our close and consistent ties with universities, with the Senate of Berlin, and with institutes of science.

Looking back at the good and the bad times, Schering has certainly achieved quite a bit in the past 100 years; but we have no inclination to sit back and wait for applause, because there is a lot more to be done. We know that conditions for research are getting more difficult.

- The need to produce innovative solutions is growing at the same pace as competition, and innovations cost more and more.
- The prices we ask for our products, which is the revenue that our research depends on, are becoming a critical issue in view of discussions on cost control in the health care sector.
- The legal terms of reference for our research are becoming so complex that our scientists need to know more and more about points of law and administration.

To put it in a nutshell, our relations with the outside world are getting complicated. This is a natural trend, since what we deal with is becoming more and more involved. But we are also very much aware of another trend that appears to be linked to the growing complexity of the subject. This is a diminishing trust on the part of the public in science in general, in biomedical research, and most particularly in such research conducted by industry. If the public stops accepting us, scientific progress will be jeopardized. This must not be allowed to happen.

This is one reason for you to be present at this different kind of ceremony. Instead of congratulatory speeches we will be hearing the concluding discussion of an international workshop run along the lines of the famous "Dahlem Workshop Model". Under the title "Bioscience $\rightleftharpoons$ Society", Dr. Silke Bernhard has organized and run this event, and I would like to take this opportunity to thank her most sincerely for all the work she has put into it. Likewise, I would like thank all the participants in the workshop for developing this theme, and also the Senate of the City of Berlin for helping to sponsor this workshop. I also owe a debt of gratitude to the members of the panel, who are about to delve into this topic. Finally, I must not fail to mention all the hard work put in by our Press and Public Relations department to make this meeting a big success.

Introductory Remarks

Before we get down to the final discussion I would like to make two points. First, the meeting will be held in English, since the issue is an international one and is to be discussed by international experts. Secondly, this meeting is to be recorded and published so that all interested parties can refer back to it at a later date.

So it is with great pleasure that I now introduce to you Professor David Roy, who will moderate the Round Table Discussion this afternoon. Dr. Roy chaired last week's workshop on "Bioscience $\rightleftharpoons$ Society", which forms the starting point for this Round Table Discussion. He holds degrees in mathematics, philosophy and theology, and is the founder and director of the Center for Bioethics at the Clinical Research Institute of Montreal. I am sure there is no one more competent to guide a discussion like ours, and I am looking forward to the intellectual challenge it poses.

Members of the Round Table Discussion. D.J. Roy standing

Introduction

David Roy:
Thank you very much indeed, Dr. Vita, for your kind introduction, and a very special word of thanks to Schering AG and to the Senate of Berlin for sponsoring two very important events. They have had the sensitivity and wisdom to bring us together here for a week-long workshop on "Bioscience ⇌ Society" and in particular for what I trust will be a remarkable afternoon.

Last Thursday afternoon some of the members of our working groups went on a tour to visit what used to be East and West Berlin and saw part of the remaining wall. When one of them came back he told me, "David, I saw two things painted on that wall. I think it applies to our work here this week. One of the phrases painted on the wall was: 'Save our Earth!'. The other phrase painted was: 'Get human!'." I think those two statements summarize the work that we have been trying to do.

Also within the past week two remarkable people from Germany, Lea Rosh and Eberhard Jäckel, received the Geschwister-Scholl-Preis. They suggested that a monument be erected to help us never ever to forget three things:

nie wegzusehen;
einzugreifen; und
solidarisch zu sein.

The monument should burn into our minds the message that we must never simply turn our eyes away, we must not remain passive and neutral, and we must overcome the distinction between "them" and "us".

It applies to our theme of science and society: that we go beyond the ancient idea that scientists are out there and simple society-like folk are over here, and we have to protect the one against the other. To move beyond that into close intercollaboration is the key theme that we pursued throughout the last week.

We asked four key questions:

We asked whether bioscience threatens human integrity, not just the survival of humans but their capacity to function as an evolving, productive, integrated whole; not only individually but as a society.

We asked, secondly, whether bioscience threatens ecological integrity, the capacity for productive collaboration of different sections of our environment and of our ecology.

We asked, thirdly, whether there is anything wrong in the relationship between bioscience and society.

We asked, finally, what actions we can undertake to improve the relationship between bioscience and society, if indeed there is something wrong with that relationship.

I think we may hear this afternoon, as we have heard during the week, that there are no simple answers to these questions, because scientific innovation increases both knowledge and power, and the increase in power is ambivalent. If it were not ambivalent we would not have to sit here and discuss it. It is ambivalent, this new power, because it is the power to do good as well as the power to do things that have never ever been done before.

That category – of the totally new things that have never been done before – may well enclose imaginable but highly uncertain and perhaps distant dangers. It may, secondly, house unintended and unforeseeable perils. And it may well house, thirdly, the possibility for distant and intended mischief.

It is because of the ambivalence – polyvalence, really – of that new power that we sit together today, as we have sat over the past twenty years at least, in country after country and time and time again, to bring complementary and different perspectives to bear on how, as one of our speakers said, we phrase the questions.

We have learned, perhaps from the recombinant DNA debates of the 1970s, that sensitivity to risks of danger and visions of possible mischief should be taken as grounds for observing the development of science and carefully watching it. They should not be taken as ground for stopping, for stalling, or for paralyzing scientific work. We have also probably learned the wisdom – indeed, the necessity – of initiating close and early collaboration between the scientists and the people, when scientific innovations hold forth the possibility of major impacts on individuals and on the community. These are the points that we will be pursuing this afternoon.

I would like briefly to introduce our four panel members, and then the first of our speakers.

Introduction

Evelyn Fox Keller is a historian and a philosopher of science. She has a Ph.D. from Harvard University, from the Department of Physics. She is now at the University of California at Berkeley and is known largely because of two books: *Reflections on Gender and Science* and *A Feeling for the Organism: The Life and Work of Barbara McClintock.*

Our second panel member, Dietrich Rössler, is professor for practical theology in Tübingen. He is also a member of the medical faculty there and teaches in the field of medical ethics. He has published *Der ganze Mensch, Die Vernunft der Religion,* and *Der Arzt zwischen Technik und Humanität.*

Our third panel member, Jeremy Cherfas, is presently the European correspondent for the journal *Science,* the journal of the American Association for the Advancement of Science. He is a biologist and writer with a longstanding interest in evolution, conservation, and molecular biology. He has produced *Man Made Life: A Genetic Engineering Primer.*

Our fourth panel member, Ernst-Ludwig Winnacker, has been professor of biochemistry at the University of Munich since 1980 and, since 1984, head of the "Genzentrum" in Munich. He is vice-president of the Deutsche Forschungsgemeinschaft and has written a textbook, *From Genes to Clones.* He is also a member of the Akademie der Naturforscher Leopoldina and was a member of the Enquete Commission of the Deutscher Bundestag on "Chancen und Risiken der Gentechnologie".

These four panel members will respond and explore with one another the key issues that each of our speakers will summarily present in his ten-minute presentation

Does Bioscience Threaten Human Integrity?

David Roy:
The first of our speakers is Maurice McGregor. Dr. McGregor has taught and pursued research in the fields of cardiology and cardiorespiratory physiology. At the Royal Victoria Hospital in Montreal he has served as Director of Cardiology and Physician-in-Chief. At McGill University he has served as the dean of the Faculty of Medicine and its Vice-Principal. He was Bethune Exchange Professor at Peking Medical College and also dean of the Faculty of Medicine at the University of Witwatersrand in South Africa. Dr. McGregor is currently the president of the Council for Evaluating Health Care Technologies of the province of Quebec. Many studies have been produced through that council, notably, and recently, one on the "High- versus low-osmolarity contrast media".

Maurice McGregor:
In Canada – as in most countries – the protocol for any experiment that involves people is reviewed by an ethics committee. Typically, these ethics committees are multidisciplinary and nonexpert bodies to which the scientist will explain his or her experiment and the probabilities of good or harm coming out of it. For this past week, in our section of the workshop, we have constituted such a group discussing four projects of our own choosing.

Like all ethics committees, we tried to define the problems and to judge what the ethical position of society should be in relation to each. We were nonexpert; I was probably the most nonexpert. Thus we have no expert opinions to report to you – only something of the thoughts and processes which we went through.

Question number one was: Do developments in bioscience require a new covenant between science and society? The answer – to my mind at least – is yes. My reasons – which, like the answer itself, are not necessarily those of my colleagues – go something like this:

Bioethical problems, to paraphrase Tristram Engelhardt, are just one facet of the general moral crisis caused by the loss of ethical convictions which we all face today in all sectors of society. This loss of moral authority started, let us say, with Martin Luther; he questioned the supreme moral authority of his day, and the loss of moral authority has accelerated ever since. At first, progress in science not only undermined our previous certainty of where we belonged in the cosmos, but offered the hope that reason alone might replace the moral authority which was lost, and might eventually disclose what is right and what is wrong. But, as you know, this hope of Enlightenment has, at least so far, proven vain. Even worse, with words like "God" and "spirit" deprived of a common shared meaning, we, in a pluralist society, are deprived of even a lexicon with which to discuss the issues.

In bioscience these problems are particularly pressing. The speed of progress has forced us to re-examine underlying assumptions we have never had to question before. For example, our ability to manipulate the human embryo forces us to reconsider the essence of life, and even to define precisely when we believe it to begin. Similarly, the need for donor organs in transplantation forces us to redefine death, and again, to establish precisely when it takes place.

Our discussions on these issues led me to a perception. After all the scientific evidence has been discussed and reviewed and debated, there lies at the root of these issues a belief – you may call it God-given conscience, or taste, or merely a "gut feeling", but it is a belief – that some things are acceptable while others are not. And since there is no overall moral authority, no universal truth that we can all identify, the best we can do in setting limits for science is, as Engelhardt says, to negotiate the best compromise we can between your belief and mine.

Of course this compromise may be completely unjustifiable by logic and therefore look absurd. For example, at present it is generally agreed that human embryo research is acceptable up to 14 days. On the 15th day it becomes unacceptable. Although this compromise may look absurd, the reality of the underlying feelings must nonetheless be acknowledged and honored. They are, as near as we can define it, the conscience of society.

Another perception that I drew from our discussions is that this social conscience, which is all we have left to guide us, is often diffuse, outdated by the advancing status of knowledge and very hard to determine. It is particularly hard for scientists to discuss with the scientifically illiterate – like me. But in the absence of universally applicable principles which we can apply to all cases, it becomes absolutely necessary to determine the social conscience

Does Bioscience Threaten Human Integrity?

on each issue as it arises. An example of how we do this daily is of course through ethics committees as they are applied to science or medicine.

The covenant under discussion, I would suggest, might be that we must acknowledge that scientists are a part of society and undertake to work with society to identify, through teaching and listening and debate, what the social conscience is in relation to each issue we face. This is no new covenant. We can surely reaffirm it.

What issues did we, in our ethics committee, discuss this week?

We discussed the dangers of the new knowledge available for genetic counseling. It was comforting to be reassured that in general, parental choice was exercised so wisely that a conflict between the accumulated results of individual decisions and the social good seemed unlikely. Some of us, however, expressed reservations and believed that this situation would require monitoring.

We discussed the thorny issues surrounding the experimental and therapeutically potentially beneficial use of fetal tissue, a matter as difficult and incapable of resolution as the issue of abortion itself. Again, negotiated compromise seemed the only civilized possibility.

But the greatest interest in our group was devoted not to high-tech bioscience, but to a question of the application of relatively low-tech interventions. And this was natural, since this was an issue affecting not only the health and lives of millions, but the integrity of mankind and the biosphere we inhabit. This discussion concerned the ethical dilemma: What should determine our response to the population explosion?

The population of the world, we were told, is increasing at a net rate of about one million every four days. The existing hypothesis has been that raising productivity and increasing educational levels will, given time, automatically result in a fall in the birth rate wherever it is high. This may or may not be true. But in many countries we will never find out, because in them the population burden is already so great that the ecosystem is absolutely not capable of sustaining increased productivity. Worse, in some instances it is in immanent danger of complete collapse. The disasters that will follow will make almost anything that we can imagine appear small.

With this background in mind, it has been asked whether it could ever be moral to promote cost-effective health interventions such as vaccination or oral rehydration on a wide scale? Could it be moral if we knew that the result, increased infant survival, will only aggravate the impending disaster?

I must tell you that the answer in our ethics committee was immediate and definite: A deliberate withdrawal of aid to the living – whatever the consequences – was totally unacceptable. Such an act would be inconsistent with human integrity. The species that might survive through the application of such a measure would survive with its humanity reduced. There was a second reaction to this issue, which was also rapid and, I think, unanimous: To omit from the health measures given to our neighbors, particularly in the Third World, the knowledge and means of population control would be inexcusable. The only measure that we can take which might mitigate the extent of the anticipated disaster would be the immediate and generous distribution of the means of population control to all who lack them. Irrespective of the overall ecological consequences, to deprive mothers of the ability to regulate the size of their families was an unacceptable deprivation of freedom.

Indeed, we felt in our group that there was an urgent need to exert mass pressure on those civil and religious authorities who were opposing the dissemination of birth control knowledge and means, and who were withdrawing support from research directed to the development of simpler, more applicable methods. We may (although this is not the suggestion of my group) hope that the great pharmaceutical companies who have the most advanced expertise in this area will direct some of their vast resources to the development of a simpler, cheaper, and more applicable contraceptive technology. No financial investment could, surely, be too great to save humanity from this disaster.

Discussion:
Does Bioscience Threaten Human Integrity?

David Roy:
The speed of progress in the biosciences, Dr. McGregor said at one point, forces us to question underlying assumptions that we really never thought we had to question before.

Members of the panel, can you comment on that?

Jeremy Cherfas:
I am concerned about the issue of drawing hard-and-fast lines when we all know that so many processes are a continuum. Dr. McGregor mentioned the 14-day limit on embryo research that has been introduced in my own country, Great Britain, and I agree with him that at the moment 14 days looks hard to justify. Why not $14\,^1/_2$? Why not $13\,^1/_2$? But a line we had to draw, and a line we have drawn.

What concerns me is that we have always drawn a line between the human species, *Homo sapiens,* and the rest of the natural world, and I am worried that we might be tempted to draw our line somewhere else. I am sure I needn't remind this audience that we share more than 99% of our DNA with our closest relatives, the chimpanzee and the gorilla. And I am wondering whether the ability to draw hard-and-fast lines between our species and other species is going to become harder and harder as we learn more about the infinitesimal differences between ourselves and other species.

Ernst-Ludwig Winnacker:
Your comment comparing us with our nearest neighbors makes an interesting point, namely that there are limits to what modern bioscience can likely achieve. When you say that we share most of our genetic information with the chimpanzee, you imply that the abilities that make us human – certain cognitive achievements – cannot have a genetic basis. And if they do not have a genetic basis, they cannot be manipulated by genetic means. The point here

is that there is a limit to what bioscience can do in certain fields. Many of the Frankenstein proposals and Frankenstein ideas – which have been with us for at least the past 150 years or so – are, really, science fiction. I think that we should, when we speak of these matters, turn the argument around and say: This really shows that there are limits to what we can accomplish in science – and in particular in genetics.

Evelyn Fox Keller:
I am tempted to ask, quite irreverently, whether you are talking about First World or Third World chimpanzees. That is to say, I shift the question. The question of the relation between humans and chimpanzees – or humans and other species – seems to me more abstract and of less immediate urgency than some other questions that Dr. McGregor put forward.

In view of the absence, or the demise, of Enlightenment ethics, and of confidence in universal moral and ethical principles, I fear that we are caught in a rather disingenuous place. We still believe in the notion of expertise, and now we are extending that expertise beyond the purely scientific. Who is to provide the ethical expertise for the very pluralist world in which we live?

We continue to rely on tried-and-true principles of liberal democracy, individual choice, parental choice. But even we – at least in the United States – are in practice somewhat disingenuous in the application of these principles. For instance, the value of parental choice seems clear when parents are of a certain sort, but when it comes to addict mothers, the question takes on an entirely different aspect. Are addict mothers entitled to make the same mistakes that, say, middle-class mothers might be entitled to make?

Even more to the point: do countries have the same rights to choose that we grant to middle-class parents? Do they have the right to make their own choices on population policies, even, say, to make their own mistakes? Why is it that we, who are trying to decide how to manage population crises impending in other countries, mouth the platitudes of choice, abjure coercion, and yet are willing to intervene, to exert, in Dr. McGregor's words, mass pressure on the opponents of policies that only we can agree on? To what extent are we actually willing to intervene in the policies of other countries, and to what extent are we willing to let them make their own choices – or even their own mistakes?

Dietrich Rössler:
I would like to come back to the statement about sharing DNA with other species. We may share DNA, but we can't share moral responsibilities, can we? And thus there is a difference in bearing responsibility for what transpires,

Does Bioscience Threaten Human Integrity?

for the world, for human life and so on. What is the point of mentioning that we share our DNA with other animals?

Cherfas:
Because I question the assumption that a bad human being is better than a good chimpanzee.

We are in danger sometimes of ignoring our responsibilities and avoiding the tough questions. That is why I mentioned it, because I think that, as we learn more and more about embryological development, we must note that it has become harder to say what is and what is not acceptable to do to a human embryo. Having faced the problem of having to decide – now we may decide that there is no answer.

Roy:
You started off with the point about the human embryo, referring back to McGregor's reference to the 14-day limit. Why do you think that it is dangerous? You didn't think that it was dangerous to draw a line there, you just didn't want it to be a hard-and-fast line, an absolute line that would never be subject to negotiation again. Is that the point?

Cherfas:
That is a subsequent point. I certainly would not want it to be a hard-and-fast line forever and for all time. We may discover that in fact 14 days is too long and we should stop at 10 days; we may discover that we want to continue beyond that. That is a subsequent point. My original statement was meant more in the context of information making it harder to see clear distinctions. Knowledge gained through the biosciences forces us to examine what we thought were simple distinctions.

Roy:
Could we focus attention for just a few minutes on this embryo research question? Why is it that when one speaks with intelligent people in several different countries one comes up with radically different policies? One country may permit research on the human embryo within certain periods of time, other countries bar it completely, and yet others say, "Well, you can do research on the embryos, but these must be embryos left over from attempts to achieve a pregnancy. You must not create embryos specifically for research purposes."? Is there a logic that explains or can help to explain these differences?

Fox Keller:
Cultural logic.

Roy:
Can you help me to understand why this takes place?

Rössler:
It characterizes a compromise, and compromises develop through environment, culture, and tradition. Compromise in Germany is something different from compromise in the United States, as we know. And compromises have to be achieved under regional conditions.

Roy:
But are there many rights and many wrongs? Is one of these countries correct, and the others simply backward?

Rössler:
The difference you point out is a consequence of the loss of ethical convictions and moral authorities that Dr. McGregor mentioned. That loss of a common cultural or moral tradition leads to the point where we have compromises which differ one from the other.

Roy:
But in such an important matter of basic, fundamental manipulative research with emerging human life, can we afford to live with radically different policies? In different parts of the world?

Cherfas:
We do live with radically different political systems in different parts of the world.

Roy:
But are political systems as dangerous as contemporary science?

Cherfas:
They control contemporary science.

Winnacker:
I think here we can all agree. But let me make one point. I agree entirely that a 14-day limit – or $13\frac{1}{2}$ or $14\frac{1}{2}$ – is absurd and has no scientific basis. In this country, however, we have a zero-day limit, and I must say that I would prefer a situation in which applications in human embryo research would be

Does Bioscience Threaten Human Integrity?

judged on their respective merits rather then being prohibited from the outset. This I believe because no questions relating to this field would be answered if there were no experimentation. I think therefore that a ban is most likely the worst thing one can do in such difficult situations.

Cherfas:
I would be interested to know whether those countries that have banned research on embryos will be willing to accept the results gained in those countries that have taken the moral risk of allowing research on embryos. Will you forgo the medical benefits to be had as a result, if there are any, which I suspect that there may be? Britain allows research on embryos and takes the moral blame for allowing it. Will you accept the benefits?

Winnacker:
You put me in a difficult position here, because I dislike legislation on science to begin with, but this question can be extended to many other fields. There are people who fight vigorously against genetic engineering and who take recombinant insulin to treat their diabetes, and rightly so. There are people adamantly opposed to using jet aircraft, who, when they have to travel ...

Fox Keller:
I am surprised that David is willing to discuss the question of a universal law or agreement on embryo research and not discuss the tremendous variation in abortion legislation, given that these two issues are so tightly related. I am struck by the avoidance of the whole question of abortion. We know that there is tremendous variation in abortion regulations. We also know that it is a particularly acute issue at this moment, in the United States and in Germany, and even more in Eastern Europe. I don't hear anybody pushing for universal abortion legislation.

Cherfas:
Would you like somebody to push for it?

Fox Keller:
No, actually not. I do believe in the unrestricted right to abortion, but to push for that right would also, in my view, be another clear example of cultural imperialism.

Rössler:
We live in different moral communities, as Tristram Engelhardt has pointed out, and we have to respect the differences of different moral communities.

Fox Keller:
Does that also include repression of certain voices in other moral communities? We speak of these notions – moral communities – as if they were homogeneous. The problem is that all of these communities are highly structured and operate on the basis of the suppression – repression – of different voices. Where do we stand with regard to that?

Roy:
A sort of civil excommunication or marginalization of those who do not articulate the dominant community view? Is that what you mean?

Rössler:
I think a moral community is a smaller population than a nation such as the United States or Germany. Moral communities are confessional communities, or feminist communities, and the one must respect the other. That is what I think of as moral behavior.

Cherfas:
Let me come back to the issue I raised: Will you allow the use of the products of somebody else's moral decision? If your moral community frowns on either embryo research or abortion, will you allow members of that community who do not share those views to take advantage of somebody else's moral community?

Roy:
The question has arisen in another form. At least in North America we would refuse to publish the results of a scientific research project or experiment – however valuable those results may be – if the experiment were basically unethical or had been conducted in a basically unethical way. We would say, "Yes, good results, but tough luck."

You seem to be posing a question that is different but related to that with respect to countries or groups who might say, "Oh, no! Embryo research. Terrible, terrible!". And then, if something very beneficial were to come out of it? You seem to be implying that if I, in that country, were morally consistent I would say, "No, I am not going to benefit from that." Is that what you are asking?

Cherfas:
That is precisely what I am asking, yes. But I do not expect moral consistency. That, I think, would be too much.

I think we can see the beginnings of, if you like, a global marketplace of ideas, just as we have in labor and in moral freedoms, and in economies. I can invest my money more or less where I want to; I can take my factories more or less where I want to; and I think we do see, on some of these questions, people taking their scientific curiosity where they will be free to exercise it. That worries me.

Roy:
So one group can rejoice that there is another group or country, much more liberal, much less scrupulous with respect to its research. We can retain our moral integrity, let them do the work that we find morally reprehensible, and then we can benefit from the results.

Fox Keller:
I am actually quite puzzled by Jeremy raising this question. It rather strikingly resembles another question that has been raised: Why should we share the technological benefits of our genome research with those countries that don't pay for it?

Cherfas:
Yes, it does. And I think I would be disinclined to support withholding the results of such research.

Fox Keller:
But you would be willing to support it in a moral domain?

Cherfas:
I think so. But you are putting me on the spot, and, as I said, I don't necessarily advocate consistency.

Roy:
Do you know how many times in this past week I found myself exactly in that position, of saying something and then not really knowing how I could defend it, or even whether or not I should?

Cherfas:
We are talking about gut feeling. One of the great things of the past week is how often one has been brought up short and said, "Well, do you feel the same way about another subject?" And I applaud Evelyn for doing that. All I can say is that I never said I was going to be consistent. I could try, but I am

not sure that consistency is what we are aiming at. What we are aiming at is a sort of humanity and a sort of integrity. But that may not be consistent.

Fox Keller:
If you do want to push this question, let me remind you of a point that came up just two minutes ago, that communities, nations certainly, are not morally homogeneous, and what you are suggesting sounds dangerously like intending to punish an entire nation for the compromises achieved by a moral majority of that nation. It would – to return to the abortion question – be like refusing abortions to those women who could not get abortions in their home country or home state.

Cherfas:
To deny nations that do not participate in sequencing the genome the information from the genome?

Fox Keller:
Or the benefits of the embryo research.

Cherfas:
I could defend my statement by referring to another topic which has been much exercising us, the difference between science and technology. Part of the way we work is that there is a freedom of knowledge. I think that everyone – all nations – ought to have access to the results of science. But the developments of those sciences will be controlled in all sorts of ways. Sharing the sequence of the genome is fine, but if you want to use a particular genetic probe that I have developed, that is another problem.

Roy:
We did also raise questions in the course of the week about the mapping and sequencing of the human genome. This seems to be one of the areas where unexamined fears and scenarios or potential apocalyptic consequences that could result from the power that will eventually be given into our hands seem to be the most difficult to tack down. The fear seems to be very far distant.

Would anyone want to comment on the reality or unreality of concerns regarding the human genome project?

Winnacker:
I think the project as such is unavoidable, because the genome is part of our anatomy, and therefore it will be solved, one way or another.

Roy:
That it will be done, is beyond question?

Winnacker:
I think it is beyond question.

But the relevant question is: if we have this information, what is it good for and what is it not good for? Can it be misused? There are some very relevant concerns here. The benefits are quite obvious and are frequently discussed. As to the concerns, there are some very legitimate ones. For example, what will this knowledge do to people who will eventually die of a disease that will appear when they are 50 years old? What good is it to an employer or an insurance company to know whether you have a certain disease or not?

However, since we are familiar with the problems, we will be able to devise solutions to them. Several would be possible, depending on the particular questions involved. It would, for instance, be possible to create legislation. We could try to protect private information, or the relationship between doctor and patient. On the other hand, there is also the argument, frequently raised, that I have the right not to know. This is always a significant argument, but if it is accepted – and I do accept this right not to know – then the next question is whether it can actually be enforced, and whether someone else may not have the right to know.

Roy:
So I would have a right to refuse knowledge of the results of a test for, say, Huntington's disease.

Winnacker:
That is fine in your case, but then the option you choose should not be turned into an option against me, who would, perhaps, want to know.

Roy:
But the "me" in your case could be my 20-year-old girlfriend, who would like to know before I get married to her.

Winnacker:
Whether she wants to know or not is her decision and not yours, I think. It is a very intimate matter.

Roy:
Would I have a right to withhold that information from her? I am not asking you to say yes or no to that. I think it is a question that may emerge.

Fox Keller:
Your original question was: Are these fears alarmist? I think that in the long range the fears are justified by what is happening already, and I think the fears are no more alarmist than the promises are exaggerated: they both will tend to be excessive. But the fact that the genome project is already playing a part in restructuring human categories is already very clear. Categories will be replaced by genetic groupings; this is not a futuristic scenario. The Human Genome Project is changing the concept of disease; that is not a futuristic scenario, either. By changing the concept of disease it is tacitly also changing the meaning of what is abnormal; that is also not a futuristic scenario. It is happening already.

Does Bioscience Threaten Ecological Integrity?

David Roy:
I would now like to introduce our second lecturer, who is going to address the question of whether bioscience threatens ecological integrity. His name is Peter Kafka, and since 1965 he has been a research scientist in relativistic astrophysics at the Max-Planck-Institut for Physics and Astrophysics at Garching near Munich. His main research areas are cosmology, black holes, neutron stars, and gravitational waves. In the past few years he has increased his participation in discussions about energy policy and other environmental questions, and his most recent publications include the books *Kernenergie – Ja oder Nein?* and *Das Grundgesetz vom Aufstieg.*

Peter Kafka:
Our group consisted mainly of ecologists, and I did not quite belong there. I would also like to apologize to the group for acting as a provoker rather than as a moderator, which may not always have made consensus easy. I should also point out that we agreed not to differentiate between bioscience and biotechnology, because at present this would not make much sense.

The question whether bioscience threatens ecological integrity cannot really be answered scientifically because there is no rigorous scientific definition of ecological integrity. This is not surprising, in view of the fact that even in mathematical complexity theory no one has been able to define what complexity really is. The reason is that a reasonable assessment of the complexity of a system would have to take into account the whole history of the system's origin. For the biosphere this would include the history of all the "accidental fluctuations" and "bifurcations" along the path of its self-organization, because the complex hierarchical web of feedback loops which we call nature was created in those infinitely many steps of trial and error.

So it is a bit simple to say that science seeks truth. What is truth in this situation in which there are infinitely many possibilities? Within these possibilities science produces a new reality, a new truth, but we do not really know how this compares with all the other possible truths which have not been constructed. And we know little about its viability.

Is there, then, any way to see whether this production of new realities leads to more integration, rather than to the disintegration of what we have? The problem is especially difficult, because integrity cannot be assessed scientifically. Only its decomposition can sometimes be judged to some degree, and this is a very important task for bioscience – to help judge the decomposition and disintegration of a complex system. In fact, bioscience and science in general have now begun to tackle this problem of recognizing and quantifying the degradation of the environment.

The group discussed many examples, not all of which I can present here. We all felt that priority in biotechnological research should be given to diagnosis. That is, wherever one sees rapid change in an ecosystem – or globally – it is best to put money first into defining it and then into monitoring it.

Much of the discussion in our group centered on the topic of whether people should try to coexist with the old kind of nature, which "managed itself". How much change should we introduce, and how rapidly? We agreed on a general principle of precaution. If change has to be introduced – and that, too, often needs to be questioned – the first attempts should be made only locally, time should be taken for learning, and the old system should be retained somewhere, so that if the change is found to be detrimental some chance for restoration will still remain. This all sounds like little more than common sense; but where are such rules followed?

We also spent considerable time on the concept of "sustainability". What does it mean? A positive definition is difficult, although indications of non-sustainability are often clear. We agreed that by sustainability we mean the integration of people into an environment of long-lived ecosystems which provide us with food, wood, fibre, energy, beauty, an emotional home, and other "resources". At the same time, this environment must be able to coexist with our "goods" and function as a "sink" for our wastes without collapsing. Since most traditional environments are disappearing quickly and don't seem to be sustainable under present or future human populations, our question became whether science can help to find the attributes of environments we want and need and then help to retain or gain these attributes. The ecologists stressed that an important aspect of sustainability seemed to be species diversity and genetic diversity within species.

Does Bioscience Threaten Ecological Integrity? 31

We could not agree on either how great or how rapid a change would have to be for it to be seen as detrimental. Nor could we decide whether self-managing systems are superior to managed ones. But there was a strong consensus that the present state of the Earth is certainly not sustainable. This may seem trivial to many people, but I was surprised to learn that the other groups did not always agree on this. Our group had the feeling that we are in a run-away instability that is approaching collapse. This we ascribed not only to population growth and growing demands, but also to activities in parts of the world which currently have relatively stable populations. Even there, the pollution of soil, water, and air is becoming worse, the extinction of species is accelerating, and now there is the threat of changing climate.

When we look back, we see that this sort of instability has been a feature of nearly all human history. Tribes are first embedded in their environment; then they develop clever management, very slowly and in complex ways; this then allows growth beyond the carrying capacity of the environment. Attempts to increase this carrying capacity by means of new and faster management lead to the loss of the old knowledge of complex management methods. Instead, some kind of "imperialism" develops, as a society attempts to save its own structures by increased exploitation of its neighborhood, until all ends in collapse.

This process has repeated itself on growing geographical scales and shrinking time scales. It has now reached the global scale and the life time of the individual. There are no more neighbors to be conquered, but we are conquering more knowledge. The question is, can bioscience help to increase the global carrying capacity?

The ecologists expressed the magnitude of the problem: humans currently exploit some 40% of terrestrial primary productivity, and perhaps 90% of the fish in the oceans. There is thus obviously not much room left for expansion.

We discussed the details of many possibilities that might help to increase carrying capacity, and we saw many ways in which biotechnology might be of value. But a stronger feeling emerged that this is not the area in which to concentrate our efforts. We felt, instead, that it might be possible and reasonable to return to a more integrated form of agriculture, using and improving old and time-tested procedures without introducing more chemicals and more biotechnology, which often create more problems than they solve. This concept naturally includes avoiding waste production and recycling those "wastes" we do produce. To achieve such a system requires us to adapt our scientific and economic thinking to local environment and local social conditions, and this cannot be achieved from the bird's-eye view of the world which science tends to adopt today.

If we managed to return to this mode of more intensive and more integrated agriculture – closer to gardening than to farming – and if we added some small-scale aquaculture, we could, perhaps, gain a factor of 2-to-3 in carrying capacity. We could thus just manage to feed the global population of about ten billion, which is already on its way and – barring vast catastrophes – cannot be prevented.

There was a vague feeling in the group (supported by many studies) that we have a chance to approach sustainability in this way and that, in Europe, it is time to start now. Some activities we already recognize are nonsustainable, and we can develop strategies to reduce them. Energy use, for example, could be reduced by 3% every year. Over fifty years – the span over which damage to the Earth's climate will make itself felt – that will reduce our energy consumption to 20% of current levels. It would still be possible to live very comfortably at that level, because at present we waste 80% of the energy we consume.

Personally, I was surprised that a group such as ours, consisting mainly of scientists, came to a consensus that it might be possible to reach sustainability and that we should therefore try to set priorities with this in mind.

Thus, after a period of pessimism, with much stress and unease, our discussion took a more optimistic turn. We realized that our chances of finding viable futures probably do not lie with ever faster tinkering with molecules and genes, but rather in more integration at higher biological and sociological levels. As scientists, we had been afraid that our general rule of precaution might bring science to a screeching halt. So it relieved our feelings to discover that ecology is not only a part of bioscience but is also, in its complexity, a higher level of bioscience, somewhere between engineering and social science.

Once we had agreed that ecological research should be given a higher priority, and that there should be more ecology in all other developments in biotechnology, we discovered that the stress usually inherent in discussions about risk assessment for the new biotechnologies was considerably diminished. The main question, which should be put to the panel, is: Does modern biotechnology offer a new dimension of risk, compared to traditional methods? In spite of many interesting arguments in this context, we felt that the question cannot really be answered. Risk assessment necessarily deals with dangers we already expect, and we discussed many of them without becoming too frightened. But it is usually the unexpected that turns out to be most important.

What, then, constitutes rational behavior when we are offered options with unknown risks? The question really is: How many new options do we want every day? One person in our group became very angry when I asked him how many new options he wanted every day, every minute, every second. Perhaps we should not arbitrarily create options at the maximum possible speed, although there are people who take this to be science's task. Their hope that "vigilance, good will, and common sense will do" is probably unjustified. If we (and the rest of the world) don't have time to judge how the old and the new fit together, mistakes will become more and more likely, and more and more serious.

In the end I had the impression that there is rising skepticism among many ecologically educated scientists (and among many people in the rest of society) with respect to the notion that Adam Smith's invisible hand will manage ecological integrity (of which, of course, human integrity is a part). The invisible hand might turn out to be a visible foot, with a cloven hoof.

Discussion:
Does Bioscience Threaten Ecological Integrity?

David Roy:
What a particularly striking closing image.

Diagnosis is one of the words Peter Kafka stressed. One of the functions of bioscience is to help us diagnose, identify and understand the symptoms of ecological disintegration. But sometimes human patients will not believe the doctor's diagnosis; they prefer to get a second, and a third opinion, hoping that the diagnosis will be different. In extreme cases they simply reject the doctor's diagnosis, in part because they don't want to accept it, in part because they don't understand it.

Do those in charge, who carry responsibility for the common good, at times manifest a similar inability to believe the diagnosis of ecological threat and disintegration, and attempt politically to turn away from it?

Ernst-Ludwig Winnacker:
Difficult question, especially if one is not responsible for the things you mentioned. For example, it was said that there are other ways in which bioscience can improve the management of our resources. There certainly are, but the question is: Are they accepted and are there other overriding issues which prevent them from being adopted?

Take, for example, bovine growth hormone. Milk production over the past 80 years has risen by 300% per cow, with the aid of traditional genetics. And now someone new comes along and raises this figure by another 20%, leading to great excitement. The product is the same, in no respect different from ordinary milk. The only difference is the major issue behind it, which is a social one: the agricultural and farming community will have to change. This is an issue that politicians prefer not to address, a difficult question that will take a generation to solve. But it would certainly relieve resources – one would need fewer cows! If we consider how much land is used in Germany for the creation of agricultural produce, we should be happy for every inch by which it could be reduced.

Roy:
My very limited mind sees at least two questions of integrity with respect to bovine growth hormone, and I am sure there are others. The first is the integrity of the cow – to which I think Jeremy Cherfas might eventually want to say a word. There is also the integrity of the market within a particular country or between countries; so there are the two sorts of integrity, and there may be others.

Evelyn Fox Keller:
It isn't only the public or the political side of things that resists an unwelcome diagnosis such as the unsustainability of our present ecological state. It is also the scientific community. The one thing that characterizes scientific communities is their extraordinary myopia: they look for short-range solutions with the techniques they have at their disposal. One lesson I draw from Peter Kafka's remarks is the necessity for an education of scientists, an ecological education of bioscientists. We really do need to understand the magnitude of the problem and to explore the complex dimensionality of these problems before jumping off with half-baked, simple solutions.

An in-depth course in ecology should be required training for every student in the sciences. Then we might have more consensus in the scientific community about diagnoses, which, then, would be more readily believed and more credible to the community outside.

Jeremy Cherfas:
That is absolutely right. We should all be answering the question: What next? or What then? as far as we possibly can. This is not something that scientists do very readily.

I would like to pick up on the question of Adam Smith's invisible hand, because Smith's invisible hand, to mix my metaphors, requires a level playing field, and part of the problem is what are glibly described as externalities. Many of the problems in ecosystems are due to not answering the question: What next?

It may, for example, pay to put fertilizers on the land to increase prices, and the cost of the fertilizers is one factor to consider. But the cost of cleaning up the groundwater is not paid by the individual farmer and is not paid by the company that makes the fertilizers. It is paid by society, by you and me, if we want clean drinking water.

One of the things we have to learn is a new way of pricing what we have always assumed to be free. If we did have that, then I think my faith in the invisible hand would be restored

Winnacker:
But we not only want clean drinking water, we also want to be fed.

Cherfas:
Quite so. But let me be specific and concrete. People have been complaining now for ten or fifteen years about the fact that in any month of the year they can buy hard red objects in the supermarkets, things that are labeled as tomatoes but taste of nothing. This is a response to public demand. We want tomatoes in our supermarkets every month of the year at a particular price. And as a consequence, they have no taste.

There are two ways to put the taste back into tomatoes. One is the biotechnological way, the fix, which is to manufacture nonsense RNA that knocks out the gene that makes tomatoes become squashy. Then you can have ripe tomatoes, and transport them, and they will stay hard. And they may be tasty, too. The other way is simply to say, "Well, it is January. Why do I want tomatoes in January, anyway? And why should I want tomatoes from California if I live in Germany?" You solve the problem of tasteless tomatoes by returning to an earlier process.

We are often in danger of forgetting that there are frequently two ways to solve this sort of problem. One is to change the product and one is to change our expectations. I don't know which way is best, I am just suggesting we remember the two.

Winnacker:
But our desires are different. You may not want to eat tomatoes in January, but I may want to, and fifty years ago I was unable to. Now I can. The tomato might not be the ultimate example here, but it is obvious that freedom of choice is something we have gained in the last hundred years through scientific developments, and I appreciate that highly.

Cherfas:
Absolutely. But while I appreciate your freedom to eat tomatoes any time you choose, I don't appreciate the tomatoes.

Dietrich Rössler:
I would like to come back to the responsibility of the scientific community, which is a very important point. The scientific community is in need of a new education in moral and ethical questions, and it must take responsibility for its members.

An example for this is the medical profession, which does have responsibility for its individual members. That responsibility should be more effective than it is now, but there is no question that the profession has responsibility for what is done in its name. This is a model for other professional scientific communities – why not for ecologists? – to take responsibility for what is done in the name of that particular science.

Cherfas:
This idea has been floated before. It has been discussed by the Society for Conservation Biology in America, partly in connection with society's requirement to fulfill certain bits of legislation, such as environmental impact statements, which very often are assessed by people who one might prefer to have a little more insight. Certainly the Society for Conservation Biology – and I think others, too – are thinking of some sort of certification scheme for ecologists.

Winnacker:
This idea has also been pursued by people in my field of molecular biology and genetics, and not only in the medical profession. Very early on in the game of genetic engineering, the scientists – who were the only persons able to do this – went public and explained that there were potential problems. And, of course, the discussion immediately got out of their hands – which is all right – and is still continuing today. We are collecting signatures in Germany, people signing that they do not want to be involved in germ line manipulation and things of that sort, and I think that this is probably much more important than legislation.

Fox Keller:
I was asking for something more than political action or political consciousness on the part of the scientific community. It seems to me that we are in urgent need of more or different kinds of education for scientists.

The workshop on bioscience and society began with the widespread and commonly held idea that what is needed is the education of the nonscientific community. It ended with increasing recognition of the inadequacy of the education of scientists themselves. That is to say, it is not only the technology that needs to be understood, it is also the complexity of the ecosystem and the complexity of the social and political systems. The need for transforming the education of scientists is first. There is an expression: Physician, heal thyself. This is a very urgent need in the scientific community, but I do not know how to accomplish it, because I know that the demand for expertise in scientific institutions is so intense and the field has become so competitive, that in practice it has become even more difficult in 1990 than it was twenty years ago, when I first started to educate.

Roy:
May I just situate what I have seen happening here for the past few minutes. The analogy is not perfect, but at least in North America over the past fifteen years, particularly in the past ten, it is a rare thing that student physicians – and even residents – can go through their education without having a fairly intense course of medical or clinical ethics. Part of the idea behind that is that a sick person's body is part of a biography, and to achieve a treatment, and even a diagnosis, you have to see the body and the biography.

I heard Peter Kafka saying something quite similar when he spoke of the history of an ecosystem, and I hear Evelyn Fox Keller suggesting that ecology could be, for a wide range of the scientific community as it is being educated, almost like medical ethics for the medical profession.

Fox Keller:
Well, it is a fist step. I would also like them to have a course in ethics and in social and political science and in a few other things. They really do desperately need a more sophisticated understanding of social structures.

Winnacker:
I think this is all fine and good – and I guess they need to know this as citizens . . .

Fox Keller:
As scientists . . .

Winnacker:
. . . and probably also as scientists. But the process of science, of scientific discovery, is very difficult to link with such an issue. Peter Kafka is a cosmologist, where we don't have a relevant issue. But even in genetics and microbiology, the process in the course of which scientific discoveries are made has to be separated from overriding issues, social consequences, and the like.

Roy:
Isn't it a bit too much to expect that every scientist is going to be a nice culturally rounded-out general systems type of thinker? Don't we want them to get on with it, let them do their work, let somebody else worry about these big matters?

Cherfas:
In many countries we ask people who would like to be doctors or lawyers to do something else first. In other words, we seem to say to them, "You are not ready yet to be a doctor. You are going to do biology, and then become a doctor." Or we say to the lawyer, "Go out and study political science, then become a lawyer." It may be that there is now so much to science, so much that we need to be aware of, including all the things Evelyn mentioned, that we need to think in terms of shifting education around a bit.

Please don't ask me who is going to pay for it, but perhaps we should think of saying, "Okay, with a bachelor's degree we are not going to expect you to be a working scientist. But we may expect you to be a human being." We can push further training after that.

Fox Keller:
I would just like to respond to Ernst-Ludwig's complaint that it is difficult conceptually to think of how to integrate these courses. I do not think so.

First of all, I have been involved all my professional life in developing such courses. The problem is finding space in the curriculum; the resistance is from the scientists, because they are obligated to their students that the students be competitive with other students. It's a race. Creating the courses is not a conceptual problem, it is an employment problem, an economic problem.

Winnacker:
There is a cash problem, and perhaps it is possible to integrate such courses into a curriculum. But I meant something different; I meant that the process in which scientific progress is made or in which scientific developments arise is separate from this process we are thinking of. If someone makes a discovery, I think that your curricula would help him or her to judge immediately, or at least more rapidly, what implications this research might have. But the process of arriving at scientific conclusions is entirely separate from that.

Fox Keller:
That is what I am objecting to.

Rössler:
You can't solve this problem by education. It needs to be organized to help the researchers to find ways to discuss what happens with their results. We have no knowledge of what is absolutely right or wrong in these areas. Not having this knowledge, we have to engage in discussions. We all have moral intuitions, differing moral intuitions it may be, and even if we are not ethicists, we are nonetheless able to discuss moral problems. No special education is necessary to discuss moral problems.

Roy:
Perhaps what is required is the insight for the scientist as for the physician, to realize that the ability to identify, to participate in the analysis of, and eventually to resolve an ethical issue arising from his or her work as a physician is part of professional intelligence. The same thing should perhaps apply to the various scientists and scientific communities, that this is part of professional and scientific intelligence, and that one should be become proud of the capacity to participate in the identification, analysis, and resolution of the ethical issues. Until that is moved forward, a little course tacked on here and there may be a good thing, but it may not be enough.

Fox Keller:
Basically, the problem I am concerned with is the disparity between the training and understanding of the world that scientists have and the authority they are given in the world. There are really only two ways to go: either increase the education of scientists so that they become more knowledgeable, more aware, and more understanding of the world; or reduce their authority and recognize them as the technicians that in effect they are. We are not willing to do either; we attribute to them more and more social authority, while their training becomes narrower and narrower.

Winnacker:
More important, perhaps, would even be to do what science is actually all about, namely to make findings public as soon as possible. It is our duty to publicize and explain the methods and goals of whatever we do in science. If this is done properly, I think that many of the issues can be addressed very early and rapidly, and then the discussions that Dietrich Rössler mentioned can begin.

What's Wrong with the Interaction Between Bioscience and Society?

David Roy:
Our third Round Table will be introduced by Tristram Engelhardt, Jr. He is professor of medicine in the Departments of Medicine, Community Medicine and Obstetrics and Gynecology and a member of the Center for Ethics, Medicine and Public Issues, both at Baylor College of Medicine in Texas. He is also professor in the Department of Philosophy at Rice University and adjunct research fellow at the Institute of Religion in Houston, Texas. Dr. Engelhardt is the editor of the *Journal of Medicine and Philosophy* and co-editor of a series of more than thirty-five books, called the "Philosophy and Medicine Book Series"; his greatest book is *The Foundations of Bioethics*.

H. Tristram Engelhardt, Jr.:
The small group in which I spent this past week was called "Issues of the Relations of Bioscience and Society". We were instructed to explore what is wrong with those relations. Despite this instruction we looked not at what was wrong, but instead tried to characterize the kinds of tension that exist between bioscience and society, realizing that these tensions are in part helpful and useful. We made an analogy with American trial law, which proceeds on the basis of an adversarial relationship between the plaintiff (or prosecuting) and defense lawyers, realizing that men and women engaged in the pursuit of science and technology have interests that need to be laid out clearly within a forceful interchange with the various societies with which they are related. It is wrong to believe that the tensions this process creates are always unhelpful; indeed, they may in fact be useful.

We also worried about using one idea of science or one idea of society. Here I pick up on the theme that Dr. McGregor advanced in addressing the problem of framing morality and public policy in a postmodern pluralist world: we should not assume that we can find a single, secular, contentfull, canonical morality. In developing science policy, we need to look to those political structures that tolerate a plurality of moral visions: limited democracies which do not presuppose a single concrete moral understanding.

Limited democracies can resolve a difficulty noted at least indirectly at this Round Table: a state is not a moral community. We have inherited from Aristotle a misguiding vision of political life. While giving moral instruction to the man who created one of the first large-scale states, Aristotle advanced as his moral and political exemplar the polis, which he described as encompassing only 50,000 to 100,000 people, a city that can be taken in within one view. None of us lives in such a homogeneous, circumscribed political structure today. Moreover, if states are not equivalent to moral communities, it will be very hard to understand what kind of protection they should give against moral harm, such as the moral harm of having someone do embryo research in one's country. It will also be hard to understand in general secular terms why anyone would be worried about germ line genetic manipulation, other than to offer protection against unconsented-to injuries. Indeed, it becomes difficult to articulate in general secular terms the nature of such harms beyond prudential concerns regarding the protection of health, life, and the environment; and prudential reasons are not enduring barriers in principle.

I must now step back from the temptation into which I have fallen, namely, of speaking regarding my own views, and speak more as a reporter from my group: I will list six sources or areas of tension between science and society that our group explored.

First, we noted a tension born of the differences between how scientists talk about knowledge and how most men and women talk about knowledge. Scientific claims are advanced in terms of probabilities of knowing truly. Scientists realize that the results of any particular study are always qualified, and that those qualifications can be reassessed – until one must act. But when one acts, one must, at least in that action, act as if matters were sufficiently settled to move from theoretical reflections to engaging the world. The difference between what it means to know truly in theoretical contexts versus what it means to know sufficiently for practical purposes reveals a gulf between the undertakings of scientists on the one hand and citizens and policymakers on the other. To decide which risks are worth what benefits, to decide whether the risks of not having a technology are greater than the risks of having that technology, requires ranking values and goals. But establishing a ranking falls beyond the competence of scientists when we consider them only as seekers of empirical knowledge. An appeal to science will not avoid the need for value choices by citizens. The appeal to science and scientists can, however, be a means for seeming to avoid moral choices under the guise of simply discovering the facts.

Second, there are different, often competing and conflicting visions of reality. Traditional, affective, animistic, teleological, non-instrumental understandings of nature and reality contrast with views of nature as something to be used and refashioned to meet the goals of persons. Scientists and those using modern technologies may see the world quite differently from those who live outside a scientific or technological world view.

Third, we noted that the great difficulty in assessing the usefulness and risk-ladenness of technology and science is that we often do not share the same ranking of costs and benefits. Even if all individuals agree that it is important in a society to protect liberty, equality, security, and prosperity, the actual ranking one gives to those interests can result in a dictatorship or an open society. So when one looks at the moral and other value costs of science, there is no uncontroversial solution. And don't forget: we began in our group with a great hesitation about there being one single, canonical sense of society, or of society's values. Large- scale states encompass numerous societies or moral communities.

Fourth, we noted that there is always some influence on research from whoever supports or facilitates the research, whether it is government, industry or philanthropic individuals. None of these sources of support is particularly free of difficulties, and no one source is privileged. Instead, one must always note – but not be amazed that – there are such influences on science by those who support it.

Fifth, some fear that the possible risks from technological solutions to human problems will be greater than the risk from not having those solutions available. Again, there is obviously no uncontroversial way to weigh these imponderables. One confronts again the different sentiments embodied in different understandings or visions of society and of nature.

Sixth, and last, we addressed at some length the problem of communication between scientists and the general public. It is not simply that scientists, including applied scientists, have views of knowledge that differ from those held by the general public. It is also that scientists, when they seek funding or other support from society, quite understandably underscore the usefulness of science. They tend not to return again and again to the risks of science. On the other hand, the media benefit from excitement and controversy. They therefore tend to accentuate risks and problems; they also often overstate the benefits of tentative findings. The public is then confused and disappointed when science cannot possibly meet the kinds of expectations raised by the media. And the scientist may become confused and disappointed that the media –

in stressing findings that excite the public and therefore attract attention to the media – place an interest in truth second to the interests of the media. From the scientist's point of view, the problem lies in the fact that popular presentations often do not add the qualifications that reasonable scientists would have made.

Media presentations can rarely carry the nuances of hesitation and uncertainty that are a part of formal communication among scientists. But this raises once again one of the difficulties of communication between scientists and society generally. In order to communicate information to a general public, one must encapsulate very complex findings, insights, and qualifications in short messages. This compression engenders misunderstandings. In addition, the apparent coldness of technological language, as well as the professional distance that scientists must often adopt in order dispassionately to study their subject matter, often suggests, falsely, that in their personal lives scientists are not fully engaged in the concerns of humans. Scientists can thus appear strange and disjointed from contemporary cultures, even though contemporary culture depends on the modern biosciences and biotechnologies.

In short, there are difficulties in the conversation between the sciences and societies, born of differences in vision, language, perspective, and interests. Yet these very difficulties invite the open and critical exploration of issues – surely a virtue in limited democracies.

Discussion:
What's Wrong with the Interaction Between Bioscience and Society?

Ernst-Ludwig Winnacker:
You said that it is regrettable that journalists overstate the benefits of scientific results. I think it is even worse when scientists overstate the benefits of their results. This is becoming more and more of a problem. Why do scientists increasingly tend to overstate the benefits? The reason is simple, namely that the structure of scientific support is all wrong. Let us take the Human Genome Project. If someone were to give $100 million to the biomolecular community to solve the human genome problem, the problem would be solved. Every possible person would try to jump on the bandwagon. However, if we gave this $100 million not to a particular project but to biomedical research in general, then the scientists would probably not need to claim how important individual projects were; they would select the important parts of a given project.

In other words, we should reconsider how important it is to have a bottom-up approach and not a top-down approach in science.

Evelyn Fox Keller:
I am very glad that it was Ernst-Ludwig who pointed out that scientists exaggeratedly emphasize science's benefits, and relieved me of the obligation to say so. There is an idea that we should simply allow scientists free rein, just support science for the sake of pure science. But ask yourself: if we are not doing it for the results, then what is the rationale for giving $100 million to biomedical research rather then to fine arts?

My point is that there has always, since the beginnings of modern science, been a rather schizophrenic attitude toward the funding of scientific research; scientists want to be funded, and they want to be left completely alone. But – particularly since World War II – the only justification for the large sums of money demanded of society has consistently been the rewards, the benefits that will be delivered. So this double agenda is built into the structure and the size of modern science; it would not exist to the same magnitude as it does today if it were not for its technological promises. We can't separate out the delivery from the support.

David Roy:
The issue of project funding is perhaps a really critical one: enormous sums of money for particular projects as opposed to the same sums of money left to a somewhat freer play of scientific curiosity and ingenuity. One of the interesting questions, which we may not be able to pursue here, is a counterfactual one, and for that reason difficult to answer: What is it that we will miss, and perhaps miss forever, that would have been done with that money, had it – and the brains that follow it – not been directed to a specific project rather than several others, or simply left open for the play of scientific curiosity? What is it that we will never discover, that might have been discovered?

Jeremy Cherfas:
Let me answer – if I can. Peter Kafka raised the notion that we are perhaps on the brink of ecological instability. If we do not study that now, there is a good chance that we are not going to be here to study anything else. We might be, but I don't want to take the risk. Supercolliders will still be buildable in a hundred years time; quasars will still be studiable. But if we don't look at the biosphere with some care right now we are going to forgo all those pleasures. So here is a clear priority, and what is interesting is that it is finally beginning to attract some rather good brains and some political will.

Dietrich Rössler:
It is clear that we have to make choices as to what is a proper object for research and what is not. But on what grounds, what are the reasons? You say it is danger.

Cherfas:
No, not danger, timeliness. Certain things will still be studiable, if we are still here to study them, in the far-distant future. But certain other things need to be studied now because our survival may depend on them. I don't have a particular view on the value of our survival, but if you are interested in studying all sorts of things about the natural world, then this question is

What's Wrong with the Interaction Bioscience/Society?

rather timely: how we can survive in the natural world to study those things that fascinate us?

Winnacker:
Your argument is dangerous. It is acceptable only if it is broadly defined. If you define it narrowly you may miss the mark. For example, in 1953, among respected academics there were two choices with respect to polio. One was to improve the iron lung. Engineers could probably have achieved this, and even designed it to be portable. But some other person, in some entirely different field, saw that he could develop a plaque test for this virus, and since then the problem of polio has vanished.

If you define your problems too narrowly, then you waste your money.

Cherfas:
"The biosphere" is hardly a narrow definition of what we should be studying.

Fox Keller:
I would support Jeremy's proposal. The scientific project that would emerge from placing survival as the first and highest priority from the start would be radically different from any scientific project that history has thus far developed. Survival is a narrow concern, but I think it should be the first and final concern. And it would lead to a complete reordering of the hierarchy of the sciences; it would lead to a complete reconception of the meaning of what is fundamental and what is applied science. Those distinctions themselves are other orderings of what the uses of science should be.

Cherfas:
I would agree, and also point out that it has been said that philosophy is the luxury of a society with enough to eat. Many, many of the themes we study – with benefits, I am not denying the benefits – are luxuries of a society with a sustainable foundation.

Rössler:
But you don't know that beforehand, do you?

Cherfas:
I think I do. For example, although I am interested in the fundamental nature of particles, I can't see that knowledge feeding people.

Fox Keller:
It is that assumption, really, that I am challenging. Why do we consider the nature of particles fundamental? Why do we consider the knowledge of genes fundamental? I would argue that that ordering of knowledge systems has to do with what we think we can do with particles and genes, and I find it extraordinarily ironic that in the history of science, physics has automatically been taken as more fundamental than the biomedical sciences. Only now that we think we can change human nature is that order beginning to reverse. But let us ask ourselves why. What kind of culture considers the study of subatomic particles more fundamental than the study of health?

Winnacker:
We should not put values to this.

Fox Keller:
But we do!

Winnacker:
You said you want to spend $100 million on buying pictures: I have nothing against that.

Fox Keller:
That was a rhetorical question. We don't in fact have the money to spend to satisfy all our interests.

Rössler:
You define benefits first, then you establish a hierarchy of values. The first step is not a scientific step, not a scientific question. It is a political question to say: we need more milk, let us have more cows.

Cherfas:
All I was suggesting was that if you are interested in preserving something here for people, if you believe you should perhaps try to save future generations somewhere to live, and live well, then to me it is self-evidently obvious that the first thing to study is the support system on which we depend. The rest are luxuries. If you do not accept that and if you do not care about what happens as a result of our activities on this Earth, then, of course, you should study anything you want to. Those are not scientific decisions.

What's Wrong with the Interaction Bioscience/Society? 51

Rössler:
It is not a question of scientific method and argument. That is what I would like to make clear.

Fox Keller:
Well, then nothing is, because there is no science that proceeds without a value, without a goal, without a benefit. Schering, I assume, supports a great deal of basic research, and it does so with the clear expectation of profit. It would not make sense to do otherwise. Almost the entire history of physics has been developed since it has been supported with the expectation of military benefits. There were always goals and values behind every scientific development.

What Jeremy is suggesting is that we put up front the goal of future human survival, a very anthropocentric position, which I embrace entirely.

Roy:
Engelhardt spoke of tensions between these two large bodies – science and society –, tensions that are potentially productive, not necessarily destructive. Any further thoughts on that?

Cherfas:
Perhaps you don't want to change direction, but I am interested in the idea of dialogue and the idea of society and scientists engaging one another. And I am particularly worried, on a very practical level, at the reluctance of some scientists to take some representatives of society seriously. This is becoming acute in biotechnology, and also in animal rights. Too many scientists, I fear, have not engaged in a dialogue with opponents of the use of animals because they feel that there is nothing they could say that might change their minds. Maybe scientists need to take on those opponents, but they also need to address themselves directly to the public, because the public is to a large extent undecided on these matters, and at present hears only one side of the story.

You may not think it worth talking to somebody who would bomb a researcher working on better anesthesia, which has happened in England. But I think it is important that scientists talk to the people and not dismiss the public because of a few self-elected representatives of that society.

Roy:
Do you think we are currently set up and prepared to achieve the kinds of ongoing communication between scientists and the general public that you are talking about? Do scientists even believe that they should be communicating?

Winnacker:
There are many scientists, including myself, who would agree that communicating with the public is important. But have you ever heard of someone given the chance to talk about a problem of bioscience and conducting a discussion like ours on prime-time television? This is considered boring, and scientists have very few opportunities to publicize their ideas, even if they want to. Since you now work as a journalist, how do you think it should be done?

Cherfas:
It is not a question of how journalists think it should be done and how scientists should do it. I think, for example, that scientists have sometimes deliberately chosen to say, "Leave me alone, let me get on with it. I take public money, but I am not going to justify what I am doing, and I am not going to explain what I am doing." This is, once again, a luxury. You will never force television to put boring people on: they will never do it. But most scientists are not boring. They must engage ordinary people at every opportunity, take the opportunities that are offered and foster a sense of responsibility. They should try not to educate but to inform and to engage with the public.

Roy:
Do scientists not fear that they might lose the respect of their own colleagues if they are seen too frequently, too publicly, too gloriously, on television and radio and in glossy newspaper articles?

Fox Keller:
I think there are two different agendas that are being confused here.

At least in the United States there is a surprising amount of science on prime-time television, because television has discovered that science attracts large audiences. But the form of presentation is hardly dialogue. The scientist comes on stage to inform the public what it is they need to know, and it is entirely up to the scientists whether information is communicated with all the qualifications Tristram Engelhardt referred to or without any qualifications whatsoever. That is an arbitrary judgement that is left in the hands of the scientists.

What we lack is any opportunity of real dialogue between scientists and other members of society with quite different interests and different perceptions. That dialogue is extremely difficult. Having talked to scientists as a scientist and as a humanist, having occupied both positions, on both sides of the university, for some time, I have discovered that it is extremely difficult

What's Wrong with the Interaction Bioscience/Society? 53

to maintain any conversation that transgresses or threatens to transgress any of their beliefs, for example the belief in the transparency of their language, or the truth or objectivity of their concerns. One of the reasons it is such an incredibly difficult enterprise is that scientists are not educated to respect other perspectives. That is a simple fact of their education. And you cannot have dialogue without respect.

Roy:
How many of us in society are educated to respect other perspectives?

Fox Keller:
For most of us, far more than for the scientific end of the campus.

Cherfas:
It depends on what you want to do with that respect. I can respect beliefs that don't try to impinge on the way I live my life.

Fox Keller:
The scientists are impinging on the way the rest of us live our lives.

Roy:
But isn't too much respect exactly what we don't want? Tristram Engelhardt mentioned forceful interchange. Forceful interchange can mean, "I think you are simply bloody wrong."

Fox Keller:
Actually, he invoked a legal model, and I want to respond to that. He also wants to evoke the postmodernist notion of pluralism: that there are no universal truths, no homogeneous moral communities, and that all that is possible is local consensus.

For that, the model of legal negotiation is entirely inappropriate, because the legal model wants to collapse all discussion. Legal discourse is designed to adjudicate between two opponents, which requires reducing a complex interplay of many different structured themes into poles. I think we want something quite different as a model, something which will allow for the textuality of the different themes and the different issues.

Roy:
How can forceful interchange be possible if one group retains a disproportionate amount of knowledge, power, and force? How can such an interchange really effectively take place in society? We can spend hours saying that scien-

tists have to communicate more effectively, in a more integrated fashion with the rest of society. Is that not in part utopian if the sciences themselves have developed and differentiated into many complex subbranches even within one particular, general scientific area such as the biosciences? How can the information that is generated along these different branches ever be sufficiently integrated to permit communication rather than a simple increase in noise?

Winnacker:
One of the answers is that communication is not the only problem in exchanges between society and scientists. Many others have been brought up here, and one that I would like to discuss is the problem of risk perception.

As a scientist, risk is the product of the probability that something could happen and the magnitude of the damage which might result. But people in general often forget the probability issue and speak only of the magnitude of the potential damage. Thus every possible organism in this world is regarded as an organism of pestilence, even though this is not true. But the fact that it is not true can often not be communicated. This is just an example, and there are many issues behind it, such as fear and the excessively rapid development of a field.

Roy:
Fear, yes, but do you think it is really possible for the general public to catch up with the galloping horse of science, which has gone so far down the road, so quickly?

Fox Keller:
I want to challenge the ordering here. We are familiar with the characterization of the public as mathematically illiterate, fearful, and incapable of distinguishing between relative magnitudes of one-in-a-million and one-in-a-billion. I want to turn the tables. What is it that prevents scientists from addressing the problem posed by the ecological crisis? What is it that prevents scientists from thinking about problems that they don't know how to handle? Is that fear, or myopia? It is not just that scientists have technical information that the public has to catch up on. People outside the scientific community have their eye on issues that it would be well for scientists to heed, yet they don't. Indeed, they seem unable to. What stops them?

Winnacker:
Sometimes scientists may even be modest; perhaps scientists realize that it is easy to say that we should work on ecological problems, but realize also that this is not so easily done.

What's Wrong with the Interaction Bioscience/Society?

Fox Keller:
Perhaps. But perhaps, too, scientists are making the same kind of mistake you accuse the public of: misperceiving risk by paying attention only to the danger and not to the probabilities. Perhaps scientists are paying attention only to the probabilities or the ease of solution, and not paying enough attention to the consequences.

Rössler:
I think the gap between these two tendencies is not a gap between the scientific community and society. It is a gap right in the midst of society itself. One tendency has the risks in view and the other has the benefits in view. There is not much hope of changing people from one side to the other. The task is mediation: to find compromises between what is possible and what is necessary. That is not a discussion between the scientific community and society. It is also a discussion within the scientific community itself. You have scientists who belong to the one tendency and others who belong to the other tendency. This means that mediation is not a task for the scientific community; it is a political task to organize possibilities and find compromises.

What Actions Are Required to Improve the Present Uneasy Relationship Between Bioscience and Society?

David Roy:
Dr. Margaret Maxey is our next speaker. She has recently become president of the newly formed National Institute of Engineering Ethics, which is affiliated with the National Society of Professional Engineers. She is currently professor of bioethics in the biomedical engineering program of the College of Engineering of the University of Texas at Austin, and was formerly a consultant for Lawrence Livermore Laboratories in California on ethical issues in radioactive waste disposal. I invite Dr. Maxey to address us on the fourth of our questions.

Margaret N. Maxey:
My remarks this evening can be mercifully brief, because the discussion that you have just heard captures the essence of problems that paralyzed the group of which I was privileged to be a member. But as Socrates long ago pointed out, the unexamined life is not worth living. Similarly, an unexamined problem is not worth solving. The particular problem before us, and the way it has been formulated, opens up avenues to a wider scope and perspective in problem formulation, enabling us to be more constructive, more imaginative, and more innovative in the solutions we propose.

Centuries have passed since Bacon pointed out that knowledge is power. Is it perhaps the case that in this day and age the locus of power par excellence – both in its political and social expressions – is bioscientific knowledge? In times past physical sciences were the avenue or instrument whereby power over natural forces was exercised. Today's society is characterized by multiple pluralisms – by cultural, moral, and political diversities. Hence, as the biosciences are in their ascendancy, their effects upon society are the more contentious because diverse interpretations have complicated their power to enter the sacred regions of birth, life itself, and death. I would suggest that those who wield forms of power through bioscience are not sufficiently aware of the political uses and abuses to which their power can be put.

After a week of intense discussion, some members of our group posed an irresistible question. Have we expended a mountain of labor only to produce a mouse? I repeat that question, not facetiously, but in order to distil three points and then pose a related question.

Initially our group wrestled with the question of what should count as the public understanding of science? Should public understanding be interpreted in such a way that the arrow of illiteracy flows only in one direction? If "the public" is scientifically illiterate, might this explain why citizens are uneasy about developments in biotechnologies and bioscience? If so, deficiencies in public understanding are to blame for people's unease.

Our group took quick exception to an understanding of scientific literacy in terms of education about empirical facts. We resisted that view for a couple of reasons. First, it seemed to imply a "quick fix", whereas the collection of verifiable data can generally take 40 to 45 years. Secondly, as one member of our group pointed out, the problem is not the inability of the public to become scientifically sophisticated; it is their unwillingness, their inertia, their complacency with the status quo. This is understandable, since technology and science have been the keys to delivering a level of material well-being and a higher quality of life for more people than ever before in recorded history.

So an emphasis on scientific literacy, our group decided, is misplaced. Scientific knowledge is not factual information. Adult questions now being raised are not questions asked by children in a classroom. Simply inserting informational education into classroom settings will not solve a deeper problem.

As a society we are wrestling with economic, social, and ethical implications of the uses and abuses of scientific knowledge, and therefore our recommendation (which after listening to the discussion may seems vacuous) was twofold. We must aim vertical efforts at many publics, not to some monolithic public on which we drop propaganda leaflets. And we must also have horizontal efforts which cross the scientific disciplines. Moreover we must begin to recognize the provisionality of scientific findings. One particular member was adamant that scientific journals should be revised so that there would be one section of articles in which provisionality had been diminished, and another section of articles (peer-reviewed, of course) which were clearly identified as more provisional.

Besides this question of having multiple channels – professional journals being only one of them – we discussed ways in which we might use multiple channels for adult education as well as at less mature levels. Look, for example, at the current success in many of our nations of experiential science museums. Consider, too, the possibilities of interactive video, the development of software programs and the ways in which computerized self-paced learning tools are now being developed.

Then, when we looked at the question of communications, we simply repeated what has already been communicated here, that scientists must learn to listen to the wishes, the fears, and the concerns of various publics, and be more caring. But we also urged members of the scientific community to develop methods and techniques – perhaps even new organizations and an institutional infrastructure – designed to optimize social understanding, particularly about the thorny issue of risk-taking and the unavoidable presence of uncertainty.

The concept of risk was the most contentious issue we discussed because, initially, at least a few of us objected to the concept that risk carries a negative burden. Risk today is a four letter word. It almost automatically conjures up dangers, harms, detriments to human beings. But is it not the case that risk is our only rite-of-passage to benefits? The reason we take risks is for the foreseen and intended benefits. If perchance expected benefits do not eventuate and instead unforeseen harms occur, these are clearly unintended and unwanted. The proper symmetry exists between benefits and detriments. Risk is a neutral concept.

Some of us also took great exception to an engineering model of the idea that risk is "probability times consequences". Ordinary people are not concerned about probabilities measured in 10 to the minus 6. They are concerned about trustworthy institutions, trustworthy regulators, and trustworthy scientists. Perhaps instead of trying to measure and quantify risk as something "out there" in a physical world, we should shift our problematic arc and start talking about risk selection as a cultural bias.

Why do affluent citizens in developed nations make a selection of risks such as Alar measured in parts per billion based on extrapolations to humans from maximum tolerable doses in rodents? Why are those same people who are so exercised about Alar's hypothetical effects on children not equally exercised about the actual harm inflicted upon children by the use of drugs such as crack and cocaine? How can we explain such disparities in risk selection? The magnitudes of the consequences are clearly polar opposites; which deserves priority attention? Is something being ignored about the social process whereby people are agitated by political activists to select certain risks for decisive public action? Should the scientific community make a more constructive contribution to this social process?

We also discussed at great length the fact that research follows the money. We compared the funding from the National Institutes of Health (NIH) in the U.S.A. with the two-tiered funding of countries such as Germany, with the Max-Planck Institutes and the Federal Ministry for Research and Technology (BMFT). In Germany, there appears to be more parliamentary control over projects funded by the BMFT, the selection of which may be governed by public opinion, whereas the Max Planck allocates funds to individuals of repute. Although the group favored the NIH model, it certainly has its share of warts. The most prominent is waste: the waste of money, the waste of time (as much as 45 years before a product emerges), and the bureaucratic rigidity. Some programs have an eternal life: they are never cut off. We also lamented the fickleness of demands and the institutionalized hypocrisy: we may fund science, but we sell technology. The tyranny of time prevented us from discussing private research funding and the market mechanisms which somehow seem not to guarantee success, and this topic seems to be an uncultivated acre.

In conclusion, I ask again whether we are sufficiently aware of the politicization of science in this day and age. Should we not factor into our considerations the realization that we are now in the final decade of the second millennium?

As we approach a new century, bear in mind that a revealing phenomenon has recently been explored. In *Century's End, the Fin de Siècle from the 990s through the 1990s,* Hillel Schwartz has researched historical records of the final decade of every century since the 990s. He finds in each an upsurge in doomsday predictions, as well as a tension between those who look forward to an apocalyptic end of the world as we know it and those who persist in some form of optimism. In the 1990s we can find evidence of the same tension between those who are pejoratively referred to as Catastrophists, and those who are derisively referred to as Cornucopians, because they advocate technological substitutability as an antidote to presumed scarcities of resources.

Is the bioscience of the 1990s going to be caught between the horns of those two options and consequently weakened by the tension between these opposing world views? Will bioscientific research serve one master or the other? Or will it instead serve an urgent and basic need: to restore public confidence in trustworthy institutions? There is ample reason to believe that these conflicting world views are actually rending apart our societies under the guise of public unease about bioscience and the social impacts of biotechnologies.

What Can Improve the Relationship Bioscience/Society? 61

We are not Athenian democracies. We inhabit representative democracies fortified with mediating structures which allow the public to have some say in the direction of scientific research and the ways in which scientific research will be used. If our social fabric becomes weakened, then those mediating structures become more vulnerable to assault and fragmentation. The biosciences have a social responsibility very different from simply imparting factual or empirical information. Responsibility for restoring confidence and public trust can be achieved by face-to-face encounters such as we have experienced in this unique workshop. If anyone in this audience had observed how seriously these eminent scientists wrestled with deep ethical issues, I believe there would be no doubt that research in the biosciences is in trustworthy hands. Would that many more people could have benefited from this privilege.

Nagging questions remain. Do we now live among affluent citizens committed to picking the eyes out of the potato of life? Are we turning into a nation of hypochondriacs? If so, then I do fear for the future. Rene Dubos once called himself a despairing optimist. To each his own. Though seasoned with realism, I remain an optimist, and I hope you will all recognize that the times in which we live do not prescribe some kind of recipe for inevitable doom, but that there is every reason to forge a covenant of trust with society.

Discussion:
What Actions Are Required to Improve the Present Uneasy Relationship Between Bioscience and Society?

David Roy:
Would one of you like to pursue the notion that the general public's inability to understand science and scientific knowledge goes beyond just scientific facts to an unwillingness, inertia, lack of interest, and lack of concern?

Jeremy Cherfas:
There is no straightforward relationship between either scientific literacy or scientific understanding and fear. Those who have almost no understanding, literacy, or appreciation of science are not very fearful because they are, in a sense, unaware of the circumstances. As they gain a little more knowledge, it becomes a dangerous thing. Perhaps as they gain even more knowledge, they become able to understand that some of their fears may have been groundless.

What worries me is not scientific literacy, it is slight scientific understanding; people need to appreciate that what science produces is almost less important than the way science goes about it. This comes back to something Evelyn was saying in the previous discussion. When science appears on prime-time television, just as when it appears on the front pages of newspapers, all too often it is only the results and not the method that appear. What we scientists hold to in science is a way of finding out about things. Part of the problem is that we have failed for various reasons to establish that what thrills us – as scientists – is that we have an exciting way of knowing about things in the world. People may have the impression that what we are really interested in is the answer, but I would suggest that what we are really interested in is how we reach that answer.

Ernst-Ludwig Winnacker:
Knowledge, or scientific literacy, is certainly not the answer. It is fine to communicate scientific results, but there is much more: the problem of dialogue. The public has to feel that it is entitled to an opinion of what is being done, that it is listened to and heard, and that its concerns are taken seriously by the scientific community. This is the real key to the acceptance of technical and scientific problems.

How can we achieve this dialogue? There may be many ways, but lectures are certainly not one of them. In Germany, for instance, we have an institution called the Enquete Commission of the Parliament, where seventeen people sit around a table, only two of whom are scientists, the rest being nonscientists – politicians, social scientists and others who ask questions. That can work well.

Roy:
The Enquete Commission is an interesting model. I find it difficult to think immediately of any country in North America and Europe that has not in the last ten years had at least one major working group on, for instance, the subject of the reproductive technologies, in vitro fertilization, surrogate motherhood, and so on. Do we have any such commissions on the biosphere and the threats to the biosphere?

Evelyn Fox Keller:
I would prefer to go back to the discussion of scientific literacy, which has always seemed to me a peculiar kind of red herring. What could scientific literacy possibly mean? I have a Ph.D. in theoretical physics; that ought to make me scientifically literate. Now, if I have to make a decision – say I go to a doctor, who finds a problem and recommends a certain procedure – I have the ability and the technical training to evaluate all the data that feed into his decision, if I can get my hands on that data. But it is virtually impossible for me to get my hands on that data. I am actually talking from a recent personal experience, where the data was almost entirely inaccessible. It is not a question of not communicating enough facts or enough about the process. It is just not possible to make a sophisticated scientific judgement if you are not enmeshed in the practice and experience of that judgement. Under those circumstances, who could be scientifically literate? A scientist can't be scientifically literate; only experts on a particular question can be literate about that particular issue.

Resorting again to the medical model, that leaves our reliance on trust. We have to have confidence, just as we have to rely on our physician's judgement because we are not in a position independently to evaluate the data. What

What Can Improve the Relationship Bioscience/Society?

does it take for us to have confidence in a physician? For me it takes the assurance that the physician is listening to me, is attentive to and has thought about the concerns that affect me.

So, to go back to Dr. Maxey's introduction, what we need is not so much scientific literacy as a demonstration on the part of the scientific community that they have made an effort to find out and to attend to the concerns of the public. That effort will in turn inspire the confidence they are asking for.

Dietrich Rössler:

I would like to turn the education question the other way around. The goal of education, for instance in the medical profession, is to set the risks low and the benefits high. That is what creates the "trustworthy" institution. Now, I am a member of society and I am educated. There is a process of education taking place every day, and I am a pupil. I was educated by the scientific community, and what I have learned is: there is no risk. But having learned that there is no risk, or very little risk, I am astonished to hear that situations can arise in which risk is a great problem. The everyday process of education has to be changed so that its goal is to make trustworthy what the professions and the scientific community teach.

Cherfas:

It is not the scientists who are misinforming you about risk. It is the politicians and those who are using the scientist's work.

In Britain we have had an outbreak of a cattle disease called bovine spongiform encephalitis. The scientific reports on BSE came to the conclusion that there is indeed an unknowable and unmeasurable risk of it moving into human beings. The first report concluded by making certain recommendations, which it said would not eliminate the risk but would reduce it even further. Unfortunately, the Ministry of Agriculture, Fisheries and Food takes the attitude that the public cannot deal with such information and has attempted to fulfil what it perceives as a public need for certainty. It said that there was absolutely no risk of certain things happening; most of them have now happened, and the ministry is continually backpedalling to reassure the public.

My view is that in this case the Ministry had the power of talking to the public, but the scientists didn't exercise their's. By the time the scientists said that although they were not certain they had done everything they could to make BSE even less of a risk to people, the public had already denounced the ministry for pretending that all was completely safe. Nothing in this life is completely safe, so why did the Ministry treat me like an idiot?

Scientists have a special responsibility. If you work for government, or you work for industry, and produce conclusions, then, I think, you should stand by them even if your paymasters are trying to say something different.

I regret that in the BSE case this was not done. I don't want to be nannied; I want to be told the straight facts, but it is not the scientists who mislead me, it is the politicians.

Fox Keller:
But there are many other examples in which scientists have also been guilty of underestimating or obscuring risks. There are examples on both sides. Nevertheless, your point is well taken, that our trust and confidence depend on our being presented with as clear and unbiased an account as is available, and not being patronized.

Roy:
One of the key points that Evelyn raised concerned the conditions for investing trust. She was talking about a doctor, but also a bit more broadly about institutions, and said that she needed to feel that the physician was listening to her, attending to what she cared about, respecting her values. That process of communication and mutual concern seems to be the absolutely inescapable route to what Margaret Maxey mentioned as the need to re-establish public confidence, not only in science but in some of the other institutions that mediate between science and politics.

Rössler:
There is one perspective to add. Trustworthiness is a very vulnerable value. It can be damaged by one mistake, and it is much easier to damage it than to foster it. It is the responsibility of the members of the scientific community to care for the trustworthiness of the institutions they work for and work in.

Roy:
If it has been damaged by one such mistake, how can it be restored?

Winnacker:
This is a difficult question to answer, as we know from the Chernobyl accident. It will take ages for the atomic energy industry to recover from this problem.

I want to come to another thing Jeremy said: he doesn't want to be nannied. But science is never 100% sure about the outcome of certain events; there is no zero-risk. Can the public live with this? Can a politician be a Cassandra all the time? In cases of public health, for example, politicians may have to make real decisions rapidly on the basis of incomplete data. They cannot wait twenty years until the first cases of the disease break out.

Cherfas:
Of course not. But let me remind you that Cassandra's curse was that she was not going to be believed although she was telling the truth. It is not merely prophesying doom, it is that the doom is real but the prophecy is not believed.

But I digress. We don't want certainty. What we want is to minimize risks and to be aware of the consequences of being wrong in both directions. Let us imagine that BSE is in fact already raging through the population. If we had done nothing, that would be a terrible thing. But if we had done something – and in actual fact BSE would not have raged through the population even if we had done nothing – we would not have lost much in putting certain precautions in place.

As a general principle it is a good idea to consider the effects of being wrong in your decisions. Leaded petrol is a very good example; maybe lead does not damage children, but if it does, and we are wrong when we say that it does not, the consequences are far greater than if it does not. So the answer is to get rid of it.

As for risks, I think we have to trust the public more when we present risks. It is no good pointing out the inconsistency of refusing to eat beef and yet driving a car. That is not germane. What you could point out is that if someone is not going to eat this beef then maybe they should also think about this unpasteurized cheese that they are so fond of. We must not mix risks from completely different walks of life, because it is too easy for people to reply that they choose the risk of driving a car, but not the risk of catching BSE from beef.

Fox Keller:
I want to pick up on another point that Ernst-Ludwig made. Not only do I think that it is appropriate to a minimal respect of the public to identify the limits of our knowledge and the incompleteness of our data; I actually think it is scientifically therapeutic to be obliged to attend constantly to the limits of our knowledge and the limits of our data.

Winnacker:
I want to came back again to the problem of risk, and to the difficult point that when we talk about risks we are always comparing apples and pears. We weigh things against one another that cannot be weighed against one

another. And we are trying to make for ourselves a world in which the risk is rather low. We have agencies that develop tests that establish that certain thing are carcinogenic or not carcinogenic, and when we hear that they are not carcinogenic we are very pleased. But if we look at the past thirty years, cancer rates themselves have not really gone down, despite these activities.

We have to be careful how we tell the public where the risks lie and where not, and I think that scientists have made mistakes in these matters as well.

Roy:
For the last few minutes we have been circling around one of the points that Tristram Engelhardt raised earlier. He mentioned the uncertainty of our knowledge and the capacity to return scientifically again and again to qualify and to requalify – until we reach a point where we have to act. Then we have to behave as though the probability of what we are doing is one. We act as if we were certain. Margaret Maxey, too, mentioned the provisionality of scientific knowledge and scientific findings. What appears to be a new discovery and quite certain in one issue of, say, *The Lancet* or the *New England Journal of Medicine*, may be brought into question two months later with a subsequent study.

Have we not been circling around the difficulty of basing politically channeled warnings to the public and decisions about what should or should not be done upon information that is in part fragmented, only partially verified and still undergoing change?

Fox Keller:
Why should we assume the responsibility of converting uncertainty into an actionable premise? It is not our responsibility.

Winnacker:
This raises the question as to who is an expert.

Roy:
It also raises the question of how the uneasy relationship between bioscience and society can be improved. One of the answers is by re-establishing confidence in the institution of science. A second answer is that the scientific community should demonstrate to the rest of us that it is indeed trustworthy.

Winnacker:
There is one more point. In the biosciences we need success. The cancer situation, for example, is getting better. We have developed therapies, but the process is very slow. In the development of genetic engineering, we are

living with the first generation of drugs, all of which were known before. They were produced from animal or microbial sources, they were more expensive, they may have been less pure. But they are the first generation of genetically engineered drugs. We can only promise the second and third generations, since they do not exist yet. In the field of AIDS, we have been very fast. The virus has only been known for six or eight years, and its structure was deciphered some five years ago, but we haven't yet solved the problem. It will take a number of years. How long will the public remain patient?

In every field of technology success is rather helpful. If the development of biosciences could show more success in solving the problems that scientists, politicians, and others want to see solved, then the situation would be much easier, but what we are doing now is therefore very dangerous. I am in favor of legislation to regulate bioscience, but there is a danger that we may kill off the first young plants before they have had a chance to grow. Only if we meet with success will biotechnology be accepted, eventually.

Fox Keller:
We just have to tailor our promises, then we don't need quite so much success. We shouldn't go for the jackpot all the time. We should pursue local kinds of success with local experiments, as Peter Kafka suggested.

Rössler:
The justification of science is not only by benefits or results. We should not forget that at the very root of our tradition is the moral conviction that to have knowledge is better than not to have knowledge. That is one of the main tenets of our common cultural tradition, and it is a value which should be respected when technical results are not readily to be expected, when we ask for patience.

Cherfas:
It might be helpful to think of science and society playing the "prisoner's dilemma" game, in which science and society can cooperate, and each will derive a benefit smaller than it would derive if there were no cooperation. But the total benefit is greater than if they do not cooperate. The prisoner's dilemma is interesting, because if you are only going to play once, you should not cooperate. But if you are going to play over and over again, the best strategy is tit-for-tat. That is, I cooperate on the first go. If my partner decides not to cooperate, then I will never cooperate again. That is an unbeatable strategy: to offer you the chance to cooperate, but if you won't take it, neither will I.

My worry is that science may have already lost the first move by breaking society's trust. In which case, in fact, the best strategy may be to go our separate ways. I don't think we have lost, so we should continue to cooperate and maybe try to point out that although we don't individually do better – scientists may have to forgo some freedom, and society may have to forgo some certainty – overall we will both do better as a result of cooperation than as a result of noncooperation.

Fox Keller:
Remember, though, that "science and society" is just a way of speaking. There can be no such thing as science and society going their separate ways, because science is part of society. It cannot exist outside society. We shouldn't get carried away with that way of speaking. We are not talking about a game between two players.

Cherfas:
I am only suggesting that thinking of the relationship between science and society as a game between two players could suggest ways of fostering their complicated relationship. It is only a heuristic.

Fox Keller:
I have trouble with that heuristic.

Roy:
Could I just throw in my experience of one group of people who do believe that the first move is gone and that going our separate ways is the only way? It goes back about fifteen years, to a meeting with the chiefs of the northern Indian tribes. I had been invited to speak about general systems theory and how the holistic view of nature is like their holistic view of nature. They laughed me out of court for even beginning to suggest that the holistic view inside science is anything like their holistic view. They said to me, "David, it is too late. You in the West started this long-term scientific adventure. We believe it is doomed to disaster. We want nothing to do with it." But I said, "Now that we have started this long-term experiment, we have no choice but to stay with it." Does that say anything to you?

Fox Keller:
We have no choice but to stay with it, and nobody has the choice of opting out because there is only one world. But I want to get back to the remark that we are committed to more, that more knowledge is better than less knowledge. That is a formulation that distorts the relationship between science

What Can Improve the Relationship Bioscience/Society?

and society. Yes, more knowledge is better, but there isn't only one kind of knowledge. More knowledges is even better than more knowledge.

What we really have to insist on, over and over again, is that it is not just a question of the public learning from the scientific community, but – since we have only one world and one life to live – it is better to learn from your friends as well.

Roy:
The time that we have is just about gone, but I have the liberty of the chair to repeat a question that was posed to me by one of the audience in the hall here. He wanted Peter Kafka to amplify his point that a 3% reduction in energy use would lead over a period of time to considerably greater reductions.

Peter Kafka:
To begin with, remember that the American Indians also made a lot of mistakes. They eradicated most of the species of larger animals in North America, and yet – following catastrophic behavior – they seem to have reached a sort of equilibrium with nature again. I think there is a chance that we might also find a new, sustainable equilibrium, if we really direct all of our activities toward that aim. That was the idea of the 3% reduction.

It is trivial to state that if one recognizes that something is wrong one must think of correcting it. How fast can one stop perpetrating nonsense? Of course it cannot be done overnight. We have realized that if we continue with our present energy consumption then within about fifty years we will have ruined the climate of the Earth. The most natural and simplest solution is radically to reduce our present energy consumption within fifty years. If each year we reduce consumption by 3% of the previous year's use, we reach 20% of today's value within fifty years. During that time the Third World could increase its energy consumption to the same level of 20% of our current energy use, and this level could be maintained with clever use of the sun.

That is such an obvious strategy that I can't see why anyone would want to discuss it; it is absolutely natural.

Roy:
You may think that that is so obvious that it is not a proper subject for discussion, but I suspect that if we had more time, which we now do not, and put that on the table, we probably would have a very necessary and a very vibrant discussion.

I now, unfortunately, have to bring this Round Table discussion to a close, with a personal summary. Evelyn responded to Dietrich's point that having knowledge is better than not having it by saying that more "knowledges", not just scientific knowledge, were needed. Over the past week, and reinforced today, I have heard the following two things that fit Evelyn's point.

First, that the process of deliberation about the necessary and the sufficient conditions for integrating scientific innovation harmoniously into the fabric of our society requires the checks and the balances of quite different and ultimately complementary perspectives. The emphasis is on differing knowledges, and the effort to give them the same power as scientific knowledge is an essential challenge of the period in which we are living.

Secondly, integrating science and society in the process of establishing goals, norms and the conditions for the exercise of new power is itself an image. It is an image of human integrity, and probably the most important source of protection. The process, if it is sustained, is self-correcting. The people in the Workshop and here at the Round Table have expressed a great deal of trust in that process, and I see that trust in the process as a defining mark of a new covenant between science and society.

I hand over now to the person who will close the afternoon, Professor Günter Stock.

Closing Remarks

Günter Stock*:
This has been a long and hopefully demanding afternoon, so I will be brief.
 Our Round Table Discussion on Bioscience ⇌ Society has now reached its close. It documents some of the results of a Dahlem Workshop Model on this topic. But at the same time it also marks the end of a major experiment which we undertook jointly.
 This experiment is one of those where you have only one chance to get it right. We can only celebrate 100 years of Schering research once, and we can do it in only one way. This kind of experiment is highly appreciated by scientists, but at the same time they are also afraid of it. I must confess that I feel considerable relief that the experiment, thanks to your commitment, has been a success. But as we all know, every successful experiment often creates the need for further experiments.
 I will comment briefly on one issue. I know that it is quite an exercise for many of us to follow discussions of this kind in English. This nevertheless demonstrates several things:

- First: complicated matters call for a complicated language. What this means for our intended dialogue between science and society is a question that has not yet been answered.
- Second: it is quite an undertaking to be an international family – especially an international company – and to make use of the same language.
- Third: language, after all, is poor. For example, this afternoon we have spoken almost exclusively about bioscience and society, or bioscience versus society. In our invitation we explicitly avoided that wording, taking instead a pictogram that expressed the integration and interaction of those two expressions. The fear has always existed that words can force us into a certain type of thinking. We are not only scientists, we are society at

---

* Member of the Board of Directors of Schering AG

the same time. I am a scientist and I am society, and, actually, in many scientific fields I am more society than scientist. As a person, I don't want to be separated off. After all, I like being both.

The discussion has clearly shown that there is no general or easy way to answer pressing ethical questions. Perhaps these issues can be resolved only in small groups, as Dietrich Rössler and Tristram Engelhardt pointed out. And any resolution that is reached might hold true only for a limited period of time. This calls for ongoing effort. But it also calls for patience and tolerance in conducting the dialogue, and I think that a patient and tolerant dialogue is superior to creating curricula at universities.

We probably need to understand that the time for clear-cut answers has gone. We have to learn, and (even more difficult) to expect, that the field requires process dynamics rather than static, solid solutions. It was therefore our hope that this meeting, with the closing Round Table Discussion, would help us to encourage this badly needed dialogue between society and the biosciences. I think that in this respect we have achieved our goal.

From Schering's point of view I might say – and this view is shared by my colleagues on the Board of Directors – that we wish to communicate openly and honestly with society as regards research in general and the biosciences in particular. We must do so, and we will do so.

But what are we to communicate? If consensus is so difficult even within expert groups, it is probably very naive to hope that the mere fact of a dialogue being conducted – without authorized answers being providing – is probably part of the answer. An honest attempt at this dialogue is probably the most relevant attitude that we can adopt

A meeting such as this ends not only with a catalogue of results and new, probably better questions, but also with a need, and even more a desire, to thank David Roy for his excellent moderation. My thanks, too, to the speakers and participants in the discussions, who adhered so faithfully to the topic and brought so much dynamism and creativity to the substance of the workshop.

My appreciation also goes out to the audience, who have supported this meeting with their vital receptiveness, being forced to remain silent for at least four hours; that is quite something.

A special word of thanks is due to the people in the background, and here I would like to repeat the words of Guiseppe Vita; the people without whose organizational and logistic talents and, probably even more importantly, without whose enthusiasm, this Round Table Discussion could not have been held. There is one particular representative to whom I would like to express gratitude, and that is Dr. Klutz-Specht and her team.

Closing Remarks

Most of all, I would like again to thank Dr. Silke Bernhard, creator of the "Dahlem Workshop Model" and organizer of this meeting, for her exceptional commitment and wealth of original ideas.

Finally, I have to thank our internal discussion group, whose name probably means nothing to most of you: "Stogras", for preparing the topics of the workshop, the results of which have been discussed today.

However important discussions on the impact of bioscience may be, however important the dialogue between bioscientists and society may be, we must not forget that we, as scientists, have to be devoted to science. We have to struggle for new scientific achievements and we know that the full recognition of the problems within science is most likely not only a hindrance to science but in many cases also an important motive to further improve these achievements.

Immanuel Kant once said: It is the wind that hinders the dove's flight that at the same time carries the dove.

So, in conclusion, a very personal word: We must not forget that science always has been and always will be a lot of fun.

Foto/CV's of Lecturers/Panel Members

Jeremy Jon Cherfas:

Dr. Cherfas was born in England. He received his B.A. and Ph.D. in the field of animal behavior.

He is a science writer, broadcaster, and photographer, Previously he was life sciences consultant and biology editor of *New Scientist* magazine, frequent contributor to many other magazines and newspapers, including *The Independent, Financial Times, La Stampa, The Guardian, The Economist,* and *BBC Wildlife,* and has worked as a radio and television journalist.

Dr. Cherfas is presently the European correspondent for *Science,* the journal of the American Association for the Advancement of Science.

His broadcasts include:
Zoo 200, Battle for Leviathan, Paradise Reclaimed (documentary on tropical ecology and conservation in Costa Rica, featuring Professor Daniel Janzen, narrated by Tom Conti), and *Nature* (he was reporter for this natural and current affairs program for BBC2 TV).

His recent publications include the following books:
Man Made Life: A Genetic Engineering Primer
Darwin Up To Date; Zoo 2000: A Look Beyond the Bars
Not Work Alone: A Cross Cultural Analysis of Activities Apparently Superfluous to Survival; The Hunting of the Whale

Dr. Cherfas was the founding editor of *KEW,* the magazine of the Friends of the Royal Botanic Gardens, Kew.

He received the Cortina Ulisse Prize for Science Writing and was Highly Commended by the Glaxo Fellowships for Science Journalism

H. Tristram Engelhardt, Jr.:

Dr. Engelhardt holds an M.D. with honors from Tulane University School of Medicine (1972) and a Ph.D. from the University of Texas at Austin (1969), where he completed his undergraduate work (1963). For the academic year 1969–1970 he was a Fulbright Graduate Fellow at Bonn University, Germany, and in 1988–1989 he was a Fellow at the Institute for Advanced Study in Berlin (Germany).

He has held appointments in Baylor and Rice since January, 1983, after leaving Georgetown University, where he was the Rosemary Kennedy Professor of the Philosophy of Medicine.

Currently he is a professor in the departments of medicine as well as in the departments of community medicine and obstetrics-gynecology, and is a member of the Center for Ethics, Medicine and Public Issues, Baylor College of Medicine. In addition, he is professor in the department of philosophy, Rice University, and Adjunct Research Fellow, Institute of Religion, Houston, Texas.

Dr. Engelhardt is editor of the *Journal of Medicine and Philosophy* and co-editor of the Philosophy and Medicine book series, with over thirty-five volumes in print. He has authored over 160 articles and chapters of books in addition to numerous book reviews and other publications. He has also been co-editor of more than twenty volumes. His most recent book is *Bioethics and Secular Humanism: The Search for a Common Morality* (Philadelphia/London: Trinity Press International/SCM Press, 1991).

Peter Kafka:

Peter Kafka was born in 1933 in Berlin, Germany.

He studied at universities in Erlangen, Berlin, and Munich, and he received a diploma in physics at Munich University in 1965.

Since 1965 he has been a research scientist in "relativistic astrophysics" at the Max Planck Institute for Astrophysics in Munich and Garching. His main research areas have been cosmology, black holes, neutron stars, and gravitational waves.

Since 1977 he has increased his participation (articles, speeches, books) in the discussion about energy policy, other environmental and political questions, and the idea of progress in general.

His recent publications include the following books:

Kernenergie – Ja oder Nein (with Heinz Maier-Leibniz), Munich 1987
Das Grundgesetz vom Aufstieg, Munich, 1989

Evelyn Fox Keller:

Dr. Fox Keller was born in New York City. She received her B.A. degree at Brandeis University, her M.A. at Radcliffe College, and her Ph.D. at Harvard University, department of physics.

She has worked at Northeastern University as a visiting professor at MIT (program in science, technology, and society) and as a Fellow at Cornell University, and in the Institute for Advanced Studies at Princeton.

Since 1988 Dr. Fox Keller has been a professor at the University of California at Berkeley, departments of rhetoric and women's studies, and co-director and founder of the Berkeley Project on Bioscience and Society.

Dr. Fox Keller has received many distinguished awards and honors, among which include the Distinguished Publication Award of the Association for Women in Psychology, the Radcliffe Graduate Medal, and the AAUW's Educational Foundation 1990 Achievement Award.

She has published many books and articles, mainly in the fields of gender and science, mathematical biology, theoretical physics, and molecular biology. Her relevant articles since 1987 are:

– Demarcating public from private values inn evolutionary discourse. *Journal of the History of Biology 21(2)*, Summer 1988: 195–211.
– Problems of radical individualism in evolutionary theory. In: *The boundaries of Humanity*, U.C. Press, 1991.
– *Sexual reproduction and the central project of evolutionary theory. Biology and Philosophy 2*, 1987: 73–86.
– One woman and her theory (original title: From individual to community: The scientific journey of Lynn Margulis). *New Scientist*, July 3, 1986: 46–50.
– Secrets of of God, Nature, and Life. In: *Models of Scientific Practice*, A. Pickering, ed. (forthcoming); also in: *History of the Human Sciences*, 1990.
– Physics and the emergence of molecular biology. *Journal of the History of*

Biology, 1990.
– Just what *is* so difficult about the concept of gender as a social category? (Response to Richars and Schuster). *Social Studies of Science 19,* 1989: 721.
– Gender and science: 1990. In: *Great Ideas Today,* Great Books Series, Encyclopedia Britannica.

Margaret M. Maxey:

Dr. Maxey recieved her B.A. degree in philosophy at the Creighton University, Omaha. She took an M.A. degree in philosophy (1963) at St. Louis University and a second M.A. degree in systematic theology (1967) at the University of San Francisco. Her Ph.D. degree in Christian ethics was completed at Union Theological Seminary, New York, in 1974.

From 1970 to 1979, Dr. Maxey was associate professor of bioethics at the University of Detroit, where she was responsible for directing student programs and research at both graduate and undergraduate levels. She has also taught philosophy, ethics, and metaphysics at Barat College, Lake Forest, Illinois, at Maryville College in St. Louis, and at the Creighton University in Omaha. Dr. Maxey is a former consultant for Lawrence Livermore Laboratory, Livermore, California, on the question of ethical issues in radioactive waste disposal. She has presented invited public testimony in Canada, the Federal Republic of Germany, the Republic of the Philippines, and the United States.

She has recently become president of the newly formed National Instutite of Engineering Ethics, affiliated with the National Society of Professional Engineers. She has been a member of the board of governors of the NIEE since 1988.

Dr. Maxey is currently director of the Clint W. Murchison Sr. Chair of Free Enterprise and professor of bioethics in the biomedical engineering program of the College of Engineering at the University of Texas at Austin. Prior to her appointment, she served as assistant director of the Energy Research Institute in Columbia, South Carolina.

Dr. Maxey is a member of the American Nuclear Society; the Institute of Society, Ethics, and the Life Sciences; and the Institute for Theological Encounter with Science and Technology, the Society for Risk Analysis, and the American Society for Quality Control.

Maurice McGregor:

Dr. McGregor was born in South Africa and graduated in medicine from the University of the Witwatersrand, Johannesburg.

After four years of military service in North Africa and Italy, he took postgraduate training in internal medicine and cardiology at the University of the Witwatersrand and at Hammersmith
Hospital, the National Heart Hospital, and the Brompton Hospital in London, England.

Since 1950 he has practiced, taught, and pursued research in the fields of cardiology and cardio-respiratory physiology. On completing his training in London he joined the staff of the University of the Witwatersrand and the Johannesburg Hospital. In 1957 he went to Canada to join the staff of the Montreal Children's Hospital, the Royal Victoria Hospital, and McGill University, where he has been to the present time.

At the Royal Victoria Hospital he has served periods as director of cardiology and as physician-in-chief and at McGill University as dean of the faculty and vice-principal (health). He has served as a Bethune Exchange Professor in Peking Medical College (1974) and has served a term as the dean of the faculty of medicine at the University of the Witwatersrand (1984-1987). Dr. McGregor is the recipient of an honorary degree from his alma mater and is a professor emeritus of McGill University. Dr. McGregor is currently president of the Conseil d'Évaluation des Technologies de la Santé du Quebec.

Foto/CV's of Lecturers/Panel Members

Dietrich Rössler:

Professor Rössler was born in Kiel, Germany. He studied theology and medicine at the University of Tübingen. After his studies, he worked as an assistant physician in a psychiatric clinic. He then served as a clergyman in a country parish and was an assistant professor at the University of Göttingen.

He is, since 1965, professor for practical theology in Tübingen. In addition, he is a member of the medical faculty there and teaches in the field of medical ethics.

His publications include the following books:
Gesetz und Geschichte; Der ganze Mensch.
Die Vernunft der Religion.

David J. Roy:

Dr. Roy holds degrees in mathematics, philosophy, and theology. He earned his Ph.D. from the Westfälische Wilhelms Universität, Münster, Germany, in 1972.

He was one of Canada's three official representatives to the Summit Nations International Meetings on Bioethics. He was part of a Working Group set up by the Royal Society of Canada to assess the impact of AIDS on Canadians, and he also sat on the Working Group on AIDS set up by the Ministry of Social Affairs of the Quebec Government.

He was coordinator, along with Prof. Bernard Dickens (Faculty of Law, University of Toronto), of the section Ethics and Law of the Vth International Conference on AIDS (Montreal, June 4–9, 1989). He was also invited to be one of the plenary lecturers at this conference.

Dr. Roy is the founder and director of the Center for Bioethics, Clinical Research Institute of Montreal. He is research professor in the Department of Medicine at the Université de Montréal and has coordinated and taught courses in medical ethics and jurisprudence in the Medical Schoos of McGill University in Montreal and the Université Laval in Quebec City.

Dr. Roy serves as consultant to doctors, health care professionals, and the government on ethical problems in medicine and clinical research, nd devotes most of his time to research on ethical issues in clinical medicine and biomedical science.

Dr. Roy is also editor-in-chief of the *Jouranl of Palliative Care* and was recently appointed Chairman of the Safety and Efficacy Review Board of the Canadian HIV Trials Network.

Foto/CV's of Lecturers/Panel Members 87

Ernst-Ludwig Winnacker:

Prof. Winnacker was born in Frankfurt am Main, Germany. He studied and received his diploma in chemistry at the ETH in Zürich. He then did his postdoctorate work at the University of California at Berkeley and at the Karolinska Institute in Stockholm.

Prof. Winnacker was assistant and visiting professor at the Institute for Genetics in Cologne, Germany. He was president of the Gesellschaft für Biologische Chemie and was a member of the Enquete Commission of the Deutscher Bundestag on Chances and Risks of Gene Technology.

Presently he is a member of the Senatorial Commission for Cancer Research of the Deutsche Forschungsgemeinschaft (DFG), vice-president of the DFG, member of the Wissenschaftlich-Technischer Beirat of the Bavarian Minister-President, and member of the Working Group on Genetic Research of the German Ministry of Research and Technology.

Since 1980 he has been professor for biochemistry and the University of Munich and since 1984 head of the Genzentrum in Munich.

Prof. Winnacker has recieved the Dozentenpreis des Fonds der Chemischen Industrie and the DECHEMA Medal of the Akademie der Naturforscher Leopoldina.

List of Participants

Silke Bernhard
Koenigsallee 35, W–1000 Berlin 33, Germany

Jeremy Jon Cherfas
Top Flat, Woodlands, Bridge Road, Leigh Woods, Bristol BS8 3PB U.K.

H. Tristram Engelhardt, Jr.
Center for Ethics, Medicine and Public Issues, Baylor College of Medicine, One Baylor Plaza, Houston, TX 77030, U.S.A.

Evelyn Fox Keller
Dept. of Rhetoric, University of California, Berkeley, CA 94720 U.S.A.

Peter Kafka
Max-Planck-Institut für Astrophysik, Karl-Schwarzschild-Str. 1, W–8046 Garching, Germany

Margaret N. Maxey
Biomedical Engineering Program, Murchison Chair of Free Enterprise, Petroleum/CPE 3.168, The University of Texas at Austin, Austin, TX 78712 U.S.A.

Maurice McGregor
Conseil d'Evaluation de Technologies de la Santé du Québec,
800, Place Victoria, Tour de la Bourse, Bureau 42.05, C.P. 215,
Montréal, Québec H4Z 1E3, Canada

Barbara Riedmüller MdA
Fechnerstr. 5, W–1000 Berlin 31, Germany

Dietrich Rössler
Engelfiedshalde 39, W–7400 Tübingen, Germany

David J. Roy
Center for Bioethics, Clinical Research Institute of Montreal,
110, Ave. des Pines Ouest, Montreal, Quebec, H2W 1RZ, Canada

Günter Stock
Schering AG, Müllerstr. 178, W–1000 Berlin 65, Germany

Giuseppe Vita
Schering AG, Müllerstr. 178, W–1000 Berlin 65, Germany

Ernst-Ludwig Winnacker
Institut für Biochemie, Universität München, Karlstr. 23,
W–8000 München, Germany,
and
Laboratorium für Molekulare Biologie – Genzentrum, Am Klopferspitz,
W–8033 Martinsried, Germany